ENERGY
AND THE
MITOCHONDRION

ENERGY
AND THE
MITOCHONDRION

David E. Green
Institute for Enzyme Research
Madison, Wisconsin

Harold Baum
Chelsea College of Science and Technology
University of London
London, England

ACADEMIC PRESS   New York and London

ACADEMIC PRESS, INC.
111 Fifth Avenue, New York, New York 10003

United Kingdom Edition published by
ACADEMIC PRESS, INC. (LONDON) LTD.
Berkeley Square House, London W1X 6BA

LIBRARY OF CONGRESS CATALOG CARD NUMBER: 69 - 18346

First Printing, January 1970

Second Printing, October 1970

PRINTED IN THE UNITED STATES OF AMERICA

FOREWORD

In 1964 I had the privilege of giving the Robbins lectures at Pomona College on "The Mitochondrial Systems of Enzymes." It was suggested by Dr. Corwin Hansch of the Department of Chemistry, that these lectures be written up in book form—a suggestion with which I was agreeable at the time. The necessities of a busy life as an active scientist made it difficult to write these lectures as planned. As the years rolled by and the subject material underwent the inevitable ferment of change, I recognized that without a colleague to share the burden and help overcome literary inertia, I would be unable to give the project the time and attention it deserved. Dr. Harold Baum, now of the Chelsea College of Science and Technology of London University, has joined me in writing, not the original lectures, but a set of lectures which I would have given in 1969. The spirit and style of the lectures have been preserved but the content and emphases reflect the passage of five crucial years of development of the subject.

DAVID E. GREEN

PREFACE

The intent of *Energy and the Mitochondrion* is to present a logical, integrated interpretation of the mitochondrion as an operational and transducing unit and as a prototype of membrane systems. For reasons of economy, clarity of image, and personal prejudice, we have restricted our presentation to the point of view to which we have been led by our experimental findings documented in the bibliographies to the various chapters in this book. We do not expect that all of our interpretations will survive the test of time, but believe that the essential concepts upon which our interpretations are founded will prove correct. We have therefore avoided the temptation to be noncommittal about what is meaningful and what is not. Notions and concepts that we believe no longer have validity have been passed over in silence. There is no shortage of contemporary treatises dealing with the historical sweep of developments in the field of mitochondriology, nor is there in the literature a shortage of alternative interpretations to those that we believe have the greatest merit.

The treatment of the subject matter in the present volume represents a complete break with tradition in that the emphasis is primarily ultrastructural and molecular. Our concept of a membrane is the foundation on which our other interpretations rest. The mitochondrion serves as an introduction to membrane systems generally and it is to this broader objective that the development of the book has been directed.

DAVID E. GREEN
September, 1969   HAROLD BAUM

CONTENTS

CHAPTER VII

UTILIZATION AND MANIPULATION OF CONFORMATIONALLY CONSERVED ENERGY

CHAPTER VIII

OTHER TRANSDUCING SYSTEMS

ENERGY
AND THE
MITOCHONDRION

CHAPTER I

BIOLOGICAL
MACHINES

Living systems are capable of a wide variety of energy transformations (see Table I.1). By energy transformation or transduction we mean any process whereby the energy leaving a system (e.g. chemical energy) differs in form from the energy entering the system (e.g. light energy). Any such transformation of energy requires a transducing device. The dynamo, electric motor, gasoline engine, steam engine, telephone, radio set, television set, and storage battery are a few of the man-made transducers with which all of us are familiar in daily life. These transducing devices are machines which translate energy from one form into another. The diaphragm in the telephone translates sound energy into electrical energy. The armature in the electric motor is the key device in translating electrical energy ultimately into mechanical energy. The overall transformation may involve several component transductions. Thus, in a radio set electromagnetic radiation is translated into electrical energy, and electrical energy in turn is translated into sound energy.

Living systems also have machines which carry out energy transformations, but the biological machines are machines with a difference. They are extremely small. Biological machines are molecular machines in the sense that the transducing elements are specialized single macromolecules or sets of macromolecules rather than gross structures. Thus, the biological machines are readily contained within the confines of single cells. Indeed, as a rule, biological energy transformations are carried out not by one machine, but by hundreds or thousands of identical machines arranged in parallel and in series within individual cellular organelles, such as the mitochondrion or the chloroplast. The biological machines are thus miniaturized, "collectivized," structured chemical systems. The transducing element

Table I.1

Energy Transformations in Living Systems

Membrane or system	Input form of energy	Output form of energy
Glycolytic system	Electron transfer[a]	Bond energy of ATP
Mitochondrion	Electron transfer[a]	Bond energy of ATP
	Electron transfer[a]	Translational work (translocation of ions across a membrane against a chemical potential gradient)
Chloroplast	Light	"Uphill" electron transfer[b]
	Electron transfer[c]	Bond energy of ATP
	Electron transfer[c]	Translational work
Electric organ	Bond energy of ATP	Flow of electrical current
Nerve membrane	Bond energy of ATP[d]	Flow of electrical current
Kidney tubule	Bond energy of ATP	Translational work
Skeletal muscle	Bond energy of ATP	Mechanical work of shortening
Retinal receptor elements	Bond energy of ATP[d]	Light-triggered flow of electrical current

[a] Transfer of electrons of negative potential supplied by organic foodstuffs.

[b] Transfer of electrons to a more negative potential.

[c] Electrons of negative potential generated by prior light-energized transduction.

[d] A distinction has to be made between the trigger mechanism which sets off the nerve impulse and the ATP-energized processes by which the nerve membrane is readied for the discharge phase. The same applies to the receptor elements of the retina.

in each biological machine is a specific macromolecule or set of macromolecules uniquely fashioned for some specific transduction. Chlorophyll in the green plant, rhodopsin in the retina of the eye, and actomyosin in muscle, are well known transducing molecules found in biological machines.

Biological machines, with one notable exception (the contractile system of muscle) are constructed according to a common architectural plan. They are contained within and are intrinsic to membrane systems. We shall be discussing membrane systems *in extenso* in a later chapter. For the moment it will be sufficient to think of a membrane as a chemoarchitectural form characterized by nesting macromolecules arranged in a two-dimensional continuum without ends. The membrane dictates the repetitive character of biological machines since a membrane contains large numbers of repeating macromolecules, each one of which may be a complete transducing unit. The membranous character of biological machines enables us to anticipate those properties which derive directly from the properties of membranes.

The principal transducing devices of cells (mitochondrion, chloroplast, endoplasmic reticulum, plasma membrane, etc.) are among the universals of living systems. This is not to say that all cells contain the same sets of transducing elements; but all devices concerned with a particular transductive process will be essentially the same in principle in whichever cell they may be present. In bacteria the structural equivalent of the mitochondrion is constructed according to the same chemoarchitectural blueprint as is the mitochondrion, and involves the same or very similar operational principles.

The intent of this book is not to describe in detail all the known transducing systems of the cell, but rather to concentrate on one transducing system, the mitochondrion (Figures I.1 and I.2), and to derive from examination of this prototype the principles and concepts which have validity for all other transducing systems.

Although the transducing systems of cells are membrane-centered (with the exception noted above), this is not to say that membranes are concerned uniquely with transductions. The mitochondrion contains a lot of enzymic baggage which is not directly related to the transductions per se. Thus, many membranes are multipurpose systems charged with a variety of catalytic and physiological functions additional to the transductions. The mitochondria of mammalian cells, for example, catalyze the elongation of fatty acids and the synthesis of phospholipids. Neither of these metabolic sequences appears to have anything to do directly with the transduction of oxidative energy into the bond energy of ATP. The cell has found it convenient to house metabolic systems as well as transducing systems within the same membrane continuum; thus, the study of chemical transformations in biological systems is willy-nilly tied up geographically, if not functionally, with the study of energy transforma-

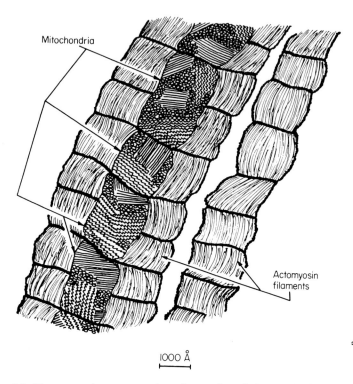

Mitochondria

Actomyosin filaments

1000 Å

Fig. I.1. Diagrammatic representation of a section of the ventricular myocardium of the canary. The four arrowheads at the left point to four contiguous mitochondria. Note the different domains of the inner membranes in the mitochondria. Based on an actual electron micrograph [see Slautterback, D. B., *J. Cell Biol.* **24**, 1 (1965), Fig. 1, p. 3].

tions. To the extent that these ancillary systems are contiguous with transducing elements, it will be necessary to refer to them in subsequent chapters. However, we are primarily concerned with presenting a bare-bones account of the essence of mitochondrial function, paying minimal attention to "optional extras," or, for that matter, to the numerous alternative descriptions of mitochondrial structure and function with which the literature abounds. This book is intended to serve as an introduction to the field, not as a comprehensive treatise.

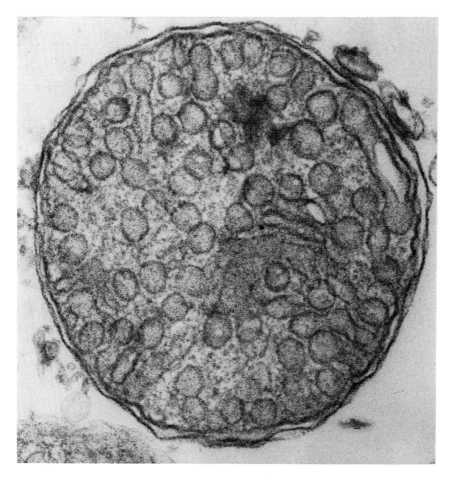

Fig. I.2. An electron micrograph of an adrenal mitochondrion. Mitochondria from different sources vary in size and form, but in any given source the size and form are distinctive. Diameter about 6000 Å.

SELECTED REFERENCES

BOOKS

Green, D. E., and Goldberger, R. F., "Molecular Insights into the Living Process," Academic Press, New York, 1967: a general reference book for biological transducing systems.

Lehninger, A. L., "The Mitochondrion," John Wiley and Sons, Inc., New York, 1964: a general reference book on the mitochondrion.

Loewy, A. G., and Siekevitz, P., "Cell Structure and Function," Holt, Rinehart and Winston, New York, 1963: a general review of cellular systems.

Mahler, H. R., and Cordes, E. H., "Biological Chemistry," Harper and Row, New York, 1966: a comprehensive reference text for general biochemistry.

SPECIAL ARTICLES

Green, D. E., and Fleischer, S., *in* "Horizons in Biochemistry" (M. Kasha and B. Pullman, eds.), Academic Press, New York, 1962: on the molecular organization of biological transducing systems.

Energy Transfer with Special Reference to Biological Systems, *in* "A General Discussion of the Faraday Society," No. 27, Aberdeen University Press, Ltd., Aberdeen, 1959: a discussion from many points of view centering on the problem of energy conservation in biological systems.

CHAPTER II

THE MITOCHONDRIAL
SYSTEMS OF
ENZYMES

In this chapter we shall be considering the overall aspects of mitochondrial functions and the interrelationships among the various functional systems, hopefully without getting sidetracked in the minutiae. In subsequent chapters each of the key systems will be reconsidered in more detail, but our immediate concern is with the role of these systems in relation to the overall workings of the mitochondrion. If it were possible to strip a mitochondrion of all but the irreducible minimal amount of equipment, such stripped mitochondria would contain a chain for the transfer of electrons (System 1), a system for generating the electron donors for the chain (System 2), and a system for synthesizing ATP by the union of ADP and inorganic phosphate (System 3). Let us consider each of these systems in turn. The essence of mitochondrial function is the transduction of free energy released by oxidation* into the energy of a new bond formed between two molecules. It is inexact to speak of the energy being concentrated in a bond, but we may proceed as though this were the case. In effect the mitochondrion accomplishes the synthesis of ATP by coupling the chemical union of inorganic phosphate and ADP

* The bonding electrons in organic molecules serving as foodstuffs are at a much higher level of potential energy than the corresponding electrons in water. Therefore, the process of photosynthesis requires solar energy to make possible the reduction of CO_2 by electrons derived from water. Thus, the flow of these electrons back to oxygen via the mitochondrial electron transfer chain consists essentially in the release of energy originating in the nuclear furnace of the sun (see Appendix I).

7

to an oxidation-reduction reaction. We may consider the synthesis and the oxidoreduction to be separate events in time and locale, coupled one to the other by a transducing device intrinsic to the electron transfer chain.

THE ELECTRON TRANSFER CHAIN (SYSTEM 1)

The electron transfer chain is a structured array of proteins containing oxidation-reduction groups which implement the stepwise (but not continuous) transfer of pairs of electrons from electron donors, such as DPNH and succinate, to molecular oxygen. Between DPNH and oxygen some twelve successive oxidoreductions take place. Between succinate and oxygen approximately the same number of oxidoreductions take place but the nature of the first three oxidoreductions is different. In point of fact, this structured array of proteins is subdivided into four units which collectively make up the electron transfer chain. These component arrays (which might be looked upon as electron transfer chains of limited span) are called Complexes I, II, III, and IV. In the oxidation of DPNH,* Complexes I, III, and IV follow one another in order. In the oxidation of succinate,* Complexes II, III, and IV provide the pathways for the flow of electrons.

Each of the four complexes is an integrated structure containing a set of three or more proteins with oxidation-reduction groups arranged in a precise sequence (see Table II.1). These are not the only proteins in the complex. In addition to the catalytic proteins, i.e. proteins with oxidation-reduction groups, there are proteins, concerned with other roles, which contain no such functional groups or at least none which are known. Each complex catalyzes the transfer of electrons between two "mobile" molecules (Figure II.1). Thus, Complex I catalyzes the oxidation of DPNH by coenzyme Q; Complex III, the oxidation of reduced coenzyme Q by cytochrome c; Complex IV, the oxidation of reduced cytochrome c by molecular oxygen; and Complex II, the oxidation of succinate by coenzyme Q. Coenzyme Q is the mobile link between Complexes I and III, and also between Complexes II

---

* DPNH and succinate are generated by the system we have designated as System 2 (see below), and are oxidized by the chain to DPN^+ and fumarate, respectively.

Table II.1

Nature of the Oxidation-Reduction Components in Each of the Four Complexes of the Chain and the Probable Sequence of Electron Flow

Complex	Oxidation-reduction components	Sequence[a,b]
I	Flavin mononucleotide (FMN) Nonheme iron (Fe)	DPNH $\rightarrow$ FMN $\rightarrow$ Fe $\dashrightarrow$ CoQ
II	Flavin adenine dinucleotide (FAD) Nonheme iron (Fe)	Succinate $\rightarrow$ FAD $\rightarrow$ Fe $\dashrightarrow$ CoQ
III	Cytochrome b (b) Cytochrome c_1 (c_1) Nonheme iron (Fe)	$CoQH_2 \rightarrow b \rightarrow c_1 \rightarrow Fe \rightarrow c$
IV	Copper$_1$ (Cu$_1$) Cytochrome a (a) Copper$_2$ (Cu$_2$) Cytochrome a_3 (a_3)	Reduced $c \rightarrow Cu_1 \rightarrow a \rightarrow Cu_2 \rightarrow a_3 \rightarrow O_2$

[a] These are the presumed sequences in which the components become reduced, but a continuous flow of electrons from one component to the next is not implied (see Chapter VI). A dotted line before an arrow indicates the probability of additional oxidation-reduction components as yet undefined.

[b] CoQ = coenzyme Q (ubiquinone); $CoQH_2$ is the reduced form of CoQ. The notations Cu_1 and Cu_2 merely indicate different species of bound copper. The cytochromes are designated only by the letters (a, a_3, b, c, and c_1) usually used to identify them.

and III; whereas cytochrome c is the mobile link between Complexes III and IV. Each of these two mobile links is reduced by one complex and oxidized by another. DPNH and succinate constitute a different category of mobiles in that they operate between the chain and a system external to the chain; they are the mobile links between the system that generates reducing equivalents on the one hand, and the electron transfer chain on the other. Finally, molecular oxygen is the

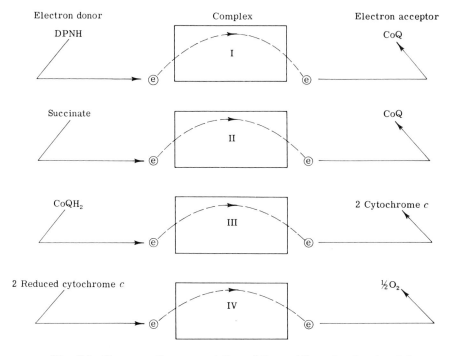

Fig. II.1. Diagrammatic representation of the mobile molecules that link complexes of the chain. Here (e) denotes a pair of electrons.

mobile link between the external environment of the mitochondrion and the electron transfer chain. Oxygen has to diffuse into the mito-chondrion from the external medium before it can accept electrons that have traversed the electron transfer chain.

As the result of the passage of electrons through Complex I, III, or IV, an energized state of the complex is generated in which the free energy liberated by the oxidoreductions is conserved as conformational energy. Thus, each of the three complexes is both an electron transfer chain and a transducing device for coupling electron flow to conformational energy. The nature of the electron transfer reaction is different in each of the three complexes, but the nature of the transduction is essentially the same, namely, the transduction of oxidative to conformational energy. The free energy conserved by this transduction is sufficient to drive the subsequent synthesis of ATP.

The potential drop between DPNH and coenzyme Q in Complex I, between reduced coenzyme Q and cytochrome c in Complex III, and between reduced cytochrome c and molecular oxygen in Complex IV, is more than sufficient to "power" in each case the synthesis of one of the pyrophosphate bonds of ATP per pair of electrons transferred (see Appendix I, Table A.1). But this is not true for the potential difference between succinate and coenzyme Q. The succinate-fumarate system has about the same oxidation-reduction potential as that of the reduced CoQ-oxidized CoQ system, and so the synthesis of ATP powered by electron flow through Complex II is energetically impossible. Thus, only Complexes I, III, and IV are concerned in coupling; Complex II is a "dummy" complex as far as coupling is concerned. Mitochondria from various sources might contain "dummy" complexes in addition to Complex II, concerned with the transfer of reducing equivalents to CoQ from other electron donors, whose oxidation-reduction potential is not sufficiently negative to reduce DPN^+. α-Glycerophosphate and possibly certain monoamines are probable candidates for the electron donors of other "dummy" complexes.

MODIFICATIONS OF SYSTEM 1

In bacterial systems, various modifications of the electron transfer chain have been found. Oxygen may be replaced by other electron acceptors, such as nitrate, sulfite, nitrogen, or nitrous oxide. These replacements require the substitution of specialized complexes for Complex IV. The electron transfer chain may be deficient in one or more complexes, as in the parasitic nematode *Ascaris lumbricoides*, which is lacking in cytochrome oxidase. In certain yeasts, Complex I can become a "dummy" complex* analogous to Complex II; such microorganisms generate only two, not three, molecules of ATP during the oxidation of one molecule of DPNH by molecular oxygen. In all cases, however, the modifications merely represent variations on a theme; no new principles of design are introduced.

GENERATION OF REDUCING EQUIVALENTS (SYSTEM 2)

Associated with the electron transfer chain in the mitochondrion is a system of enzymes that generates DPNH and succinate, the two electron donors for the chain. This system of enzymes catalyzes the metabolic sequence known as the Krebs citric acid cycle (or tricarboxylic acid cycle) (Figure II.2). Pyruvate (derived primarily from glucose via glycolysis elsewhere in the cell) is oxidized in five successive steps: pyruvate to acetyl-CoA (DPN^+ reduced); isocitrate to α-ketoglutarate (DPN^+ or TPN^+ reduced); α-ketoglutarate to succinate (DPN^+ reduced); succinate to fumarate; and finally malate to oxaloacetate (DPN^+ reduced). In addition to these five oxidoreductions there is a hydration (fumarate to malate), an isomerization (citrate to isocitrate), and a condensation (oxaloacetate + acetyl-CoA to citrate). At the end of each cycle, one molecule of pyruvate is completely oxidized,† and one molecule of oxaloacetate is regenerated, ready for another turn of the cycle (initiated by the condensation of oxaloacetate with acetyl-CoA). It will be noted that DPN^+ (or TPN^+) is reduced to DPNH (or TPNH) in four of the five oxidations of the citric acid cycle whereas succinate is generated in only one of the five oxidative steps (oxidation of α-ketoglutarate).‡ It will also be noted that one oxidative step in the cycle (succinate → fumarate) *directly*

* It would not be for reasons pertaining to oxidation-reduction potential that Complex I would be incapable of coupling electron flow to synthesis of ATP.

† Strictly speaking, the CO_2 evolved during one cycle does not originate exclusively from the molecule of pyruvate undergoing oxidation.

‡ The oxidation of α-ketoglutarate to succinate results not only in the reduction of DPN^+, but also in the generation of a bond between succinate and coenzyme A, the anhydride bond of succinyl-CoA, which is eventually translated into a bond between ADP and inorganic phosphate ($ADP-P_i$) with the release of unesterified succinate. This "substrate level phosphorylation" is analogous to the generation of ATP in the more primitive glycolytic pathway. The mechanism of this phosphorylative reaction will be discussed in Chapter VII.

involves an oxidation catalyzed by the electron transfer chain, whereas the others only generate substrates (DPNH and succinate) for the chain.

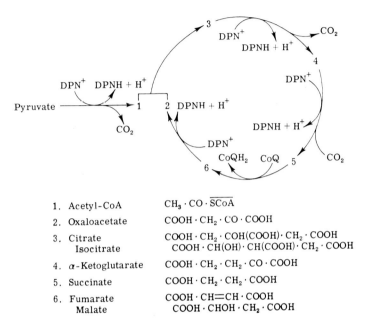

1. Acetyl-CoA $CH_3 \cdot CO \cdot \overline{SCoA}$
2. Oxaloacetate $COOH \cdot CH_2 \cdot CO \cdot COOH$
3. Citrate $COOH \cdot CH_2 \cdot COH(COOH) \cdot CH_2 \cdot COOH$
   Isocitrate $COOH \cdot CH(OH) \cdot CH(COOH) \cdot CH_2 \cdot COOH$
4. α-Ketoglutarate $COOH \cdot CH_2 \cdot CH_2 \cdot CO \cdot COOH$
5. Succinate $COOH \cdot CH_2 \cdot CH_2 \cdot COOH$
6. Fumarate $COOH \cdot CH{=}CH \cdot COOH$
   Malate $COOH \cdot CHOH \cdot CH_2 \cdot COOH$

Fig. II.2. The citric acid cycle. For simplicity, the formulae of the nonionized acids have been given. At physiological pH, the reactants are all at least partially ionized and are therefore properly referred to as the corresponding anions. The oxidation of succinate to fumarate, (5) to (6), involves the direct participation of Complex II of the electron transfer chain in the reduction of coenzyme Q.

Any molecule which can give rise to an intermediate of the citric acid cycle can in fact serve to generate succinate and DPNH. Fatty acids and amino acids in many mitochondria are used in just that capacity. When fatty acids are oxidized by β-oxidation, reducing equivalents in the form of DPNH and reduced flavin,* are generated

* Flavin eventually donates electrons to the chain, probably at the level of coenzyme Q. The full details of the reoxidation of such flavoproteins have yet to be established.

during the degradative process. From the standpoint of overall mitochondrial function, however, the relevant fact to be noted is that a molecule of acetyl-CoA is formed for each two carbon atoms in the fatty acid chain. Given a catalytic amount of oxaloacetate, this acetyl-CoA, derived from fatty acids, can readily generate both of the donors for the electron transfer chain, i.e. succinate and DPNH. Several of the amino acids can give rise to intermediates of the citric cycle by oxidative deamination (e.g. glutamate and proline are converted to α-ketoglutarate) or by transamination (transfer of the amino group to an acceptor α-keto acid whereby, for example, alanine gives rise to pyruvate and aspartate gives rise to oxaloacetate).

If intermediates in the cycle are lost by being diverted into other pathways, then the rate of generation of oxaloacetate is decreased, and this decrease in turn lowers the efficiency with which the cycle can take up acetyl-CoA units derived either from pyruvate or from the oxidation of fatty acids. An important source of intermediates to replenish the cycle, and thus restore its efficiency, is pyruvate. Pyruvate may be reductively carboxylated by reaction with CO_2 and TPNH in the presence of the "malic enzyme" giving rise to malate. It may also be directly carboxylated (by pyruvate carboxylase) to oxaloacetate, a reaction energized by ATP and probably far more important in the maintenance of the cycle than that catalyzed by the malic enzyme.

SYNTHESIS OF ATP (SYSTEM 3)

The energized state generated by electron transfer in the complex has to be "manipulated" to power the synthesis of ATP by the union of ADP and orthophosphate. This manipulation involves a synchronized discharge of the energized state and synthesis of ATP. Conformational and electrostatic energy pay the energy bill for the synthesis of ATP. This second transduction involves a system additional to, but intimately associated with, the complexes of the electron transfer chain. Once ATP is synthesized in the mitochondrion, there has to be a delivery of a phosphoryl group from internally formed ATP to ADP external to the mitochondrion. This delivery system, although different from the system that transduces conformational and electrostatic energy to the bond energy of a pyrophosphate group in ATP, may be considered a part of System 3 for translating the energized state into "available" ATP. In later chapters we shall be elaborating on the parts of this System 3.

OPTIONAL EXTRA SYSTEMS

With this brief survey, we have completed our first tour of inspection of the primary mitochondrial systems, and now we shall turn our attention to the optional systems found in association with mitochondria from various sources. In mammalian mitochondria there is a system of enzymes that can elongate fatty acids by addition of acetyl-CoA and subsequent reduction of the keto acid produced. Thus, palmitate is elongated to stearate; myristate to palmitate, etc. Some of the enzymes involved in the elongation process may also participate in β-oxidation. The exact nature of the enzymes functioning in the mitochondrial process of fatty acid elongation has yet to be clarified. Most mitochondria also contain the set of enzymes that can carry out the synthesis of lecithin and phosphatidylethanolamine from fatty acids, glycerol, and the appropriate nitrogenous base. Cytidine diphosphocholine and cytidine diphosphoethanolamine serve as intermediates in these two synthetic processes.

Why the systems for fatty acid elongation and for phospholipid synthesis are present in the mitochondrion is difficult to explain. There appears to be no direct role for these two systems in the economy of the mitochondrion, and it is possible that their association with the mitochondrion is merely incidental. One could speculate about a role for lipid biosynthesis in the assembly of the mitochondrial membranes, but until we know more about the assembly process, there is no way of assessing this possibility. The fatty acid oxidizing system is uniquely localized in the mitochondrion, but fatty acid elongation is also achieved in certain organelles other than the mitochondrion.

In addition to the contributions made by these optional systems to cell metabolism, all mitochondria participate in the intermediary metabolism of the cell, primarily through intermediates of the Krebs cycle. For example, citrate generated in the mitochondrion can eventually give rise to acetyl-CoA in the external medium. In addition, internally generated acetyl-CoA can serve as the source of acyl groups for the acylation of external CoA (carnitine is required for such transmembrane acylation reactions). In this way, the mitochondrion can influence the process of fatty acid biosynthesis external to the mitochondrion—a process which requires the availability of acetyl-CoA. Another example of the role of the mitochondrion in overall metabolism is provided by the use to which malate can be put. Malate, formed in the cycle, can diffuse out of the mitochondrion and serve as a source of reducing power and as a source of carbon in extramitochondrial gluconeogenesis.

UTILIZATION OF THE ENERGY OF THE ENERGIZED STATE

There are multiple ways in which the coupling mechanism and the energized state can be utilized by mitochondria. It can be used for the synthesis of ATP (the universal modality) but in addition it can energize translocation of ions, change in volume (ion-dependent swelling), reversed electron flow, and transhydrogenation between DPNH and TPN^+. Energy is conserved in the synthesis of ATP but is dissipated in energized swelling, in translocation of divalent metal ions, and in transhydrogenation.

The mitochondrion can concentrate divalent and monovalent cations within its membranes, and this accumulation can be powered by the energized state. The accumulation is a consequence of the interaction of salts with repeating units of the inner mitochondrial membrane, conformationally perturbed by electron flow or by hydrolysis of ATP (see Chapter VI). This interaction underlies the phenomenon of energized swelling induced by monovalent cations and also the phenomenon of the energized translocation of divalent metal ions (Me^{2+}) and inorganic phosphate with deposition of $Me_3(PO_4)_2$ in the interior of the mitochondrion. For each pair of electrons moving through a single complex, 2 divalent ions and 1.33 phosphate ions are translocated across the inner membrane into the interior space.

Among other options the movement of electrons through the chain can be coupled to the synthesis of ATP. The coupled synthesis is a reversible process. Thus, ATP can be used to reverse the direction of flow of electrons; the splitting of ATP into ADP and orthophosphate can be coupled to the generation of the energized state, and the discharge of this state in turn can be coupled to the flow of electrons in the direction opposite to the direction in which the flow is spontaneous. Such a result could have been anticipated. Maximal energy conservation requires that the coupling reactions should be carried out under reversible conditions.

One of the complexes of the mitochondrion catalyzes the hydride ion transfer represented by the equation $TPNH + DPN^+ \rightleftharpoons TPN^+ + DPNH$. The potentials of the DPNH–DPN^+ and TPNH–TPN^+ systems are virtually identical (E_0' at pH 7.0 of -0.321 V and -0.324 V, respectively), It is not surprising, therefore, that the equilibrium constant for this reaction is close to 1.0; indeed this is demanded by elementary principles of thermodynamics (see Appendix I). What is surprising is that the approach to equilibrium from right to left, i.e. with TPNH as the hydride ion donor, is much more rapid than the reverse reaction, i.e. the hydrogenation of TPN^+ by DPNH. When the mitochondrion is converted to the energized state, however, only the latter reaction takes place; moreover, the *equilibrium* is shifted far to the left.

INTERRELATIONSHIP OF SYSTEMS

Now that we have completed our brief survey of the major systems contained in the mitochondrion, it may be of value to think in terms of the interrelationships among these various systems. The links between the electron transfer chain and the citric cycle enzymes are succinate and DPNH. (For the sake of simplifying the diagram in Figure II.3, the "return" products, DPN^+ and fumarate are omitted although required to keep the cycle operative.) The enzymes oxidizing fatty acids and amino acids may be looked upon as supplementary systems that supply intermediates which feed into the citric cycle—the ultimate source of the electron donors for the chain. In bacteria the fuel for the electron transfer chain may be provided by a wide variety of molecules other than succinate or DPNH. Of course, provision has to be made for such electron donors to be utilized by the chain. This provision takes the form of complexes specialized for the oxidation of the molecules which serve as electron donors.

The passage of a pair of electrons from DPNH through the chain involves the ultimate generation of three molecules of ATP (see

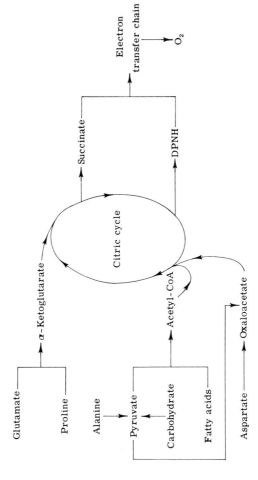

Fig. II.3. Interrelationships of mitochondrial systems.

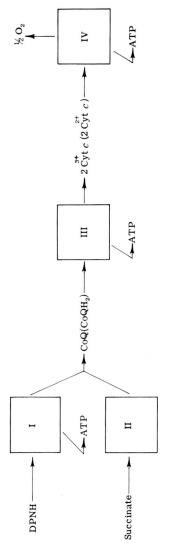

Fig. II.4. The number of coupling sites in the electron transfer chain covering the span from DPNH to oxygen, and the number of coupling sites in the electron transfer chain covering the span from succinate to oxygen.

Figure II.4). Each of the three complexes couples electron flow to
the generation of an energized state. In the oxidation of succinate, the
energized state (and hence ATP) is formed only in two complexes
during the transfer of electrons to molecular oxygen (see Figure II.4).
As we have mentioned above, Complex II is a "dummy" complex
in this respect, being incapable of generating the energized state
during electron flow.

Both the elongation of fatty acids and the synthesis of phospholipid
require ATP; this requirement is a link between the electron transfer
chain and these two synthetic systems. Moreover, there is a link
between the Krebs cycle and fatty acid elongation, in that acetyl-CoA,

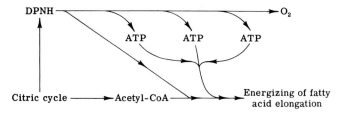

Fig. II.5. Interrelationships between fatty acid elongation, the citric cycle,
and electron transfer.

DPNH, and TPNH generated in the Krebs cycle* are utilized for
the elongation sequence (see Figure II.5). There is also a link between
the coupled synthesis of ATP and the initiation of fatty acid oxidation.
Fatty acids must be converted to their acyl-CoA esters before they
can undergo β-oxidation. This conversion is an ATP- (or GTP-)
requiring process. Thus, although β-oxidation, once initiated, can
generate ATP, this initiation depends upon the ATP-energized
conversion of the fatty acid to the acyl-CoA ester. A similar consideration applies to maintenance of the whole Krebs cycle. ATP is
needed from time to time to generate oxaloacetate from pyruvate (the

---

* TPNH may also be generated at this level of intermediary metabolism by the
reversal of the pyruvate $\rightarrow$ malate reaction catalyzed by the malic enzyme and
referred to above. TPNH may also be generated in the chain by energized hydrogenation of TPN$^+$ by DPNH. It is actually a moot point whether the isocitrate
dehydrogenase reaction in the mitochondrion can serve to generate significant
amounts of TPNH.

pyruvate carboxylase reaction mentioned above); catalytic amounts of oxaloacetate, once available, permit the oxidation of large amounts of pyruvate and the generation of much ATP.

To round out the interrelationships we must consider the various uses to which the energized state can be put. The diagram in Figure II.6 summarizes the multiple modes of disposition of the energized

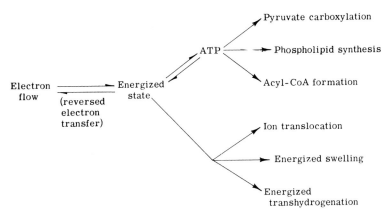

Fig. II.6 The multiple modes of utilization of the energized state.

state generated by electron transfer. There are four known options for the way in which the energized state can be put to use. When and to what extent any of the options is selected is determined by a variety of factors intrinsic and extrinsic to the mitochondrion.

SELECTED REFERENCES

BOOKS

Campbell, P. N., and Greville, G. D., eds., "Essays in Biochemistry," Vol. 1, Academic Press, New York, 1965. See essay on oxidative phosphorylation by D. E. Griffiths.

Chance, B., ed., "Energy Linked Functions of Mitochondria," Academic Press, New York, 1963.

Green, D. E., and Goldberger, R. F., eds., "Molecular Insights into the Living Process," Chapter 15, Academic Press, New York, 1967.

King, T. E., Mason, H. S., and Morrison, M., eds., "Oxidases and Related Redox Systems," Vol. II, John Wiley and Sons, Inc., New York, 1965.

Okunuki, K., Kamen, M. D., and Sekuzu, I., eds., "Structure and Functions of Cytochromes," University of Tokyo Press, Tokyo, 1968.

Slater, E. C., Kaninga, Z., and Wojtczak, L., eds., "Biochemistry of Mitochondria," Academic Press, New York, 1967.

SPECIAL ARTICLES

Chance, B., and Williams, G. R., *Advan. Enzymol.* **17**, 65 (1956): the respiratory chain.

Green, D. E., *in* "Comprehensive Biochemistry" (M. Florkin and E. H. Stotz, eds.), Vol. 14, p. 309, Elsevier Publishing Co., New York, 1966.

Green, D. E., and Allmann, D. W., *in* "Metabolic Pathways" (D. Greenberg, ed.), 3rd ed., Vol. 2, Academic Press, New York, 1968: fatty acid oxidation and synthesis.

Green, D. E., and Hatefi, Y., *Science* **133**, 3445 (1961): the mitochondrion and biochemical machines.

Green, D. E., and MacLennan, D. H., *in* "Metabolic Pathways" (D. Greenberg, ed.), 3rd ed., Vol. 1, Academic Press, New York, 1968: the mitochondrial electron transfer chain.

Green, D. E., and Silman, H. I., *Ann. Rev. Plant Physiol.* **18**, 147 (1967): the mitochondrial electron transfer chain.

Green, D. E., and Tzagoloff, A., *Arch. Biochem. Biophys.* **116**, 293 (1966): the mitochondrial electron transfer chain.

Hatefi, Y., Haavik, A. G., Fowler, L. R., and Griffiths, D. E., *J. Biol. Chem.* **237**, 2661 (1962): on the reconstitution of the electron transfer system.

Lowenstein, J. M., *in* "Metabolic Pathways" (D. Greenberg, ed.), 3rd ed., Vol. 1, Academic Press, New York, 1968: the tricarboxylic acid cycle.

KEY REFERENCES FOR CHAPTER II

Electron transfer chain

Chance (1956), Green (1966), Green and Hatefi (1961), Green and Mac-Lennan (1968), Green and Silman (1967), Green and Tzagoloff (1966), Hatefi *et al.* (1962)

System for generating reducing equivalents

Green (1966), Green and MacLennan (1968), Lowenstein (1968), Slater *et al.* (1967)

Modification of the electron transfer chain

Green and Goldberger (1967)

Utilization of the energy of the energized state

Chance (1963)

CHAPTER III

COMPOSITION AND ULTRASTRUCTURE OF MEMBRANE SYSTEMS

A membrane, in our usage, is a continuous structured layer of limited thickness enclosing a space. A vesicular membrane may be likened to a hollow ball; a tubular membrane to an inflated section of dialysis tubing. There is only one structured layer in a biological membrane (the rest is fluid-filled space), and this layer is made up of nesting particles that fit together closely to form a two-dimensional continuum (cf. Figure III.1). By our definition there are no open

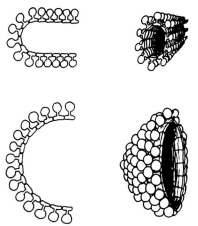

Fig. III.1. Two diagrammatic representations of a membrane as a fused continuum of repeating units. *Above:* a tubular membrane; *Below:* a vesicular membrane.

ends in a membrane. Membranes form a continuous surface like that
of a hollow sphere.

REPEATING UNITS OF MEMBRANES

Each membrane has its own distinctive repeating units—distinctive
in respect to size, composition, function, and geometric shape. This
is not to say that all the repeating units of a given membrane are
identical; there may be multiple species, but the membrane-forming
sectors of all such species of repeating units have similar sizes and
shapes. This is mandatory because all the repeating units of a given

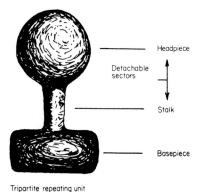

Tripartite repeating unit

Fig. III.2. Sectors of a tripartite repeating unit.

membrane must nest together, and nesting requires, if not complete
identity of size and shape of the repeating units, at least sufficient
similarity or complementarity not to compromise the nesting process.

A membrane is an expression of the properties of its repeating units
and of the mode of bonding between them. It is to the repeating units
that we must look for an understanding of what a membrane is and
how it functions.

All the membranes which couple electron transfer to synthesis of
ATP have a specialized repeating unit with three sectors (see Figures
III.2 and III.3). We shall refer to this type of repeating unit as a macrotripartite repeating unit to distinguish it from another tripartite
repeating unit of smaller dimensions which is known as a microtripartite
repeating unit. The macrotripartite repeating unit is made up of a

Fig. III.3. Electron micrographs of tripartite repeating units. A: The sarcoplasmic reticulum. B and C: The cristael membrane of the mitochondrion.

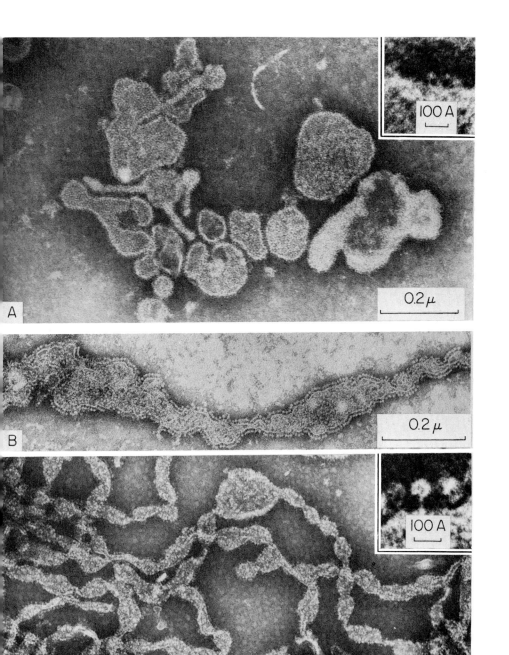

100 A

0.2μ

A

0.2μ

B

100 A

0.3μ

C

basepiece connected to a *headpiece* by a cylindrical *stalk*. The basepiece is the sector concerned in the two-dimensional nesting of repeating units. Both the stalk and headpiece can be detached from the basepiece without compromising the gross membrane structure. This is not to say that the detachable sectors of the repeating units do not influence membrane properties. There is, indeed, such an influence, but the nesting of basepieces does not require the attachment of the detachable sectors.

Each of the three sectors of the macrotripartite repeating unit fulfills a unique role in the phenomenon of coupled synthesis of ATP. We shall be considering in detail the role of each of the three sectors in later chapters. But the special point to be made is that a repeating unit must be looked upon as the molecular instrument of membrane function. In other words the design of the macrotripartite repeating unit is relevant to the coupling function of the mitochondrial and chloroplastic inner membrane. The mechanism of oxidative and photosynthetic phosphorylation is implicit in the ultrastructural pattern of the repeating unit. Thus, there is an invariant correlation between the capability of a membrane to carry out coupled synthesis of ATP and the presence of macrotripartite repeating units.

A wide variety of membranes from various sources have been found to contain tripartite repeating units of the micro variety (see Figure III.3). The characteristic of all membranes with such repeating units is the capacity to carry out ATP-energized active transport. The sarcoplasmic reticulum translocates Ca^{2+}; the plasma membrane of the red blood cell translocates K^+ and Na^+; the microvilli transport sugars and amino acids; the parietal cells of the gastric mucosa secrete H^+ into the lumen; the plasma membrane of the cells which line the gills of fish translocate Na^+ and K^+, etc. The dimensions of the microtripartite repeating units are virtually identical regardless of the source of the membrane where the repeating units are found. From this identity of dimensions and the correlation with a transport function, it is obvious that the ultrastructural design of the microtripartite repeating unit is relevant to the mechanism of active transport. As yet the precise role of each of the three sectors of the microtripartite repeating unit has not been established. Some evidence of T. Oda would suggest that the ATPase function is housed in the basepiece sector of the microtripartite repeating unit. By contrast, the ATPase function of the macrotripartite repeating unit is localized in the headpiece.

The headpiece and stalk of the microtripartite repeating units are much more tightly attached to the basepiece than are the corresponding sectors of the macrotripartite repeating units. Proteolytic digestion is the only method yet developed for detaching the headpiece-stalk sector of the microtripartite repeating units.

The class of myxoviruses have a membranous boundary layer enclosing the nucleic acid (see Figure III.4). The boundary layer

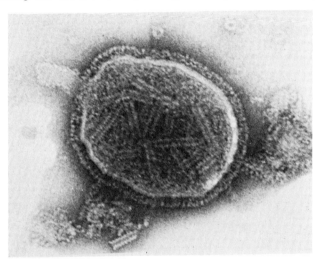

Fig. III.4. Electron micrograph of parainfluenza (Sendai) virus; negatively stained with phosphotungstate. Magnification ×225,000. The thickness of the membrane is about 70 Å; the diameter of the headpiece is 30 Å; the stalk is 13.5 Å wide and 70 Å long; the center to center distance between headpieces is 50 Å. This electron micrograph was kindly provided by Dr. E. A. Eckert of the University of Michigan School of Public Health.

corresponds in all respects to a membrane. The repeating units of the myxoviral membrane also have a tripartite character. The dimensions more closely approximate those of the microtripartite repeating units than those of the macrotripartite repeating units. Present indications are that the repeating units of the membrane of the myxoviruses are specialized for invasiveness. The membrane can be stripped away from the nucleic acid network in the interior of the virus, and used instead of the whole virus for developing immunity in man against the viral agents. One of the three sectors of the viral tripartite repeating unit has a hemoagglutinin which is detachable from the membrane. This detachability would argue for the localization of the agglutinin in the headpiece-stalk sector of the repeating unit, i.e. in the detachable sectors. Yet another sector contains neuraminidase—an enzyme which hydrolyzes the ester link between sialic acid and mucoprotein.

The fact that the membrane of all the viruses of the myxovirus class have virtually identical tripartite repeating units would be consistent with the hypothesis that the viral tripartite repeating units are specialized for the task of attachment to and penetration of the appropriate membranes of the host organism. We would not expect, therefore, to find the functions of the various sectors of the viral tripartite repeating units to be the same as those of the macro- or microtripartite repeating units.

The repeating units of biological membranes fall into one of two categories—multipartite and monopartite. We have already considered the examples of multipartite repeating units, such as the macro-, micro-, and viral tripartite repeating units. The outer membrane of the mitochondrion and the plasma membrane of the erythrocyte have repeating units without projecting sectors. These repeating units correspond to the basepieces of the tripartite repeating units. The designation monopartite refers to the repeating units in membranes which do not show projecting sectors.

Some membranes have only one type of repeating unit while other membranes have more than one type. The inner membrane of the mitochondrion has exclusively macrotripartite repeating units; the membrane of the sarcoplasmic reticulum has exclusively microtripartite repeating units. However, the plasma membrane of the red blood corpuscle has at least two types of repeating units—monopartite and microtripartite. Red blood cells with high capability for concentration of K^+ show a relatively large number of microtripartite repeating units whereas red blood cells with a low capability for concentrating K^+ show a relatively small number of such repeating units.

The repeating unit is not a loose or random association of the component sectors. The detachable sector in the crista of the intact mitochondrion is always in a 1:1 relation to the basepiece, and the stalk is always in a 1:1 relation to the headpiece or basepiece. Thus, different sectors are integral parts of a repeating unit, and there are precise chemical links (albeit not always strong ones) which bind the sectors together.

The sectors of a repeating unit were first discovered and visualized by electron microscopy, but now it is possible to separate by chemical means the repeating unit of a given membrane into its component sectors and to study each of the sectors in isolation.

COMPOSITION OF REPEATING UNITS

The repeating units of the mitochondrial cristael membrane are built up from molecules of protein and lipid bonded together according to a precise blueprint. Each sector has its characteristic set of specific protein molecules. Moreover, the lipid is not randomly distributed within the repeating unit; it is localized at certain sites, and nowhere else. The repeating unit of the inner mitochondrial membrane has a molecular weight of about 8 to 9×10^5; the estimates of the molecular weights of basepiece, stalk, and headpiece, respectively are 5×10^5, 0.25×10^5, and 3×10^5. Protein accounts for about 70% of the total mass and phospholipid for the rest. On the basis of the analytical data it would appear that a basepiece of the inner mitochondrial membrane contains no less than 2 molecules of a noncatalytic protein (molecular weight about 60,000), about 5 molecules of catalytic protein (molecular weight about 25,000), and some 300 molecules of phospholipid. It is probable that lipid is associated largely, if not exclusively, with the basepiece sectors of repeating units.

BIMODAL NATURE OF PHOSPHOLIPID

The determinant of membrane formation, i.e. the determinant of the way in which particles line up one layer thick to form a continuum, is phospholipid. We must, therefore, consider those features of the phospholipid molecule which account for this determination of membrane structure. Phospholipids are bimodal (amphipathic) molecules. The hydrocarbon chains of the fatty acid residues make up the nonpolar sector of the molecule; the nitrogenous base plus the phosphate residue make up the polar sectors. The geometry of the phospholipid molecule is such that the nonpolar and polar sectors are at opposite ends of the molecule. Monomeric phospholipid molecules of natural origin (with fatty acid residues of 16 carbon atoms or more) have essentially zero solubility in water. The molecules, however, tend to align themselves so that the nonpolar sectors are directed away from the aqueous solvent* and the polar sectors are directed into the aqueous solvent. One arrangement allowing such molecules to satisfy this tendency is that of an air–water interface (lowering of surface tension is an index of such alignment) or that of an oil–water interface (the detergent action of phospholipids is a manifestation of this arrangement). Another arrangement is that of a polymeric micelle. This micellar arrangement makes it possible for arrays of phospholipid molecules (as distinguished from individual molecules) to be soluble in water (cf. Figure III.5). The micelle can assume various forms and shapes (rosettes, bimolecular leaflets, etc.). In micelles, the nonpolar sectors of the phospholipid molecules are bonded hydrophobically to one another whereas the polar sectors are directed into the aqueous medium.

BINDING BETWEEN PHOSPHOLIPID AND PROTEIN

The phospholipid molecules associated with the basepieces of the repeating units are hydrophobically bonded to protein. Titration studies reveal that all the negatively charged groups of phospholipid molecules bonded to protein in a repeating unit are available for interaction with basic proteins such as cytochrome c or basic poly-amino acids, such as polylysine. It has been difficult to rationalize the nature of the hydrophobic binding of phospholipid to membrane proteins. Recent studies of Perutz on aqueous channels in hemo-globin have suggested a rational basis for such binding. As shown in Figure III.6, the repeating units are assumed to be honeycombed

---

* Nonpolar chains make thermodynamically unfavorable holes in the aqueous medium. Maximal hydrogen bonding between water molecules in the neighbor-hood of a nonpolar chain can be achieved only at the expense of a considerable ordering of the water structure. The resulting decrease in entropy is particularly unfavorable at higher temperatures.

Fig. III.5. Electron micrographs of phospholipid micelles in water. (A) large field; (B) "exploded" micelle; and (C) compressed micelle.

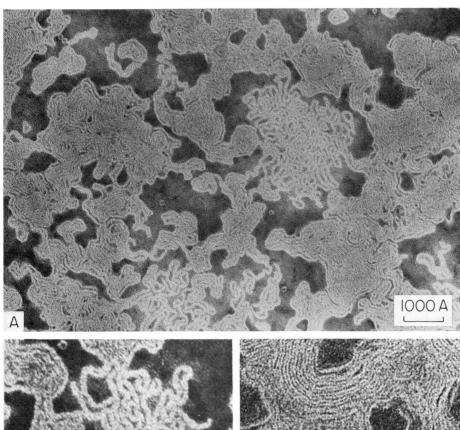

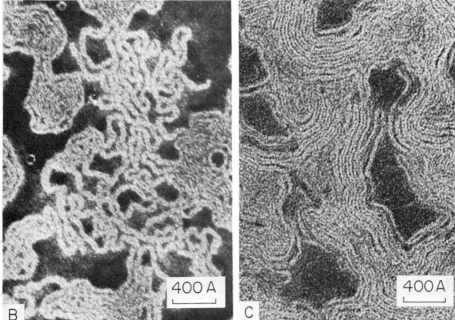

with cavities or channels between the component proteins. In hemo-globin, the cavities are lined with polar amino acids and are filled with water. If the interprotein spaces in a membrane were lined with nonpolar amino acids, this would favor the filling of these spaces with the aliphatic tails of phospholipids rather than with water. All, or nearly all, of the lipid could be taken up in this way if the average spacing of proteins in membranes is similar to that in crystals of proteins such as cytochrome c or lysozyme. Penetration of the aliphatic chains of the phospholipid molecules into the interprotein

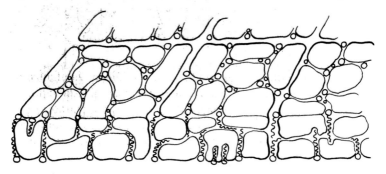

Fig. III.6. Channels between nesting proteins in the membrane-forming sectors in which phospholipid molecules can fit.

Two assumptions are implicit in this formulation of the binding of phospholipid to protein in the membrane: (1) that the channels are at right angles to the direction of the membrane, and (2) that the amino acid residues lining these channels have a nonpolar character. The fatty acid residues of the phospholipid molecules that fit into these channels are bonded hydrophobically to the amino acid residues of the proteins that line the channel. Some of the channels may accommodate only one phospholipid molecule; others may accommodate several molecules. This formulation has been proposed by G. Vanderkooi.

spaces should be possible, just as it is possible for small molecules to diffuse into a protein crystal, since the chains are flexible, and a CH_2 group is about the same size as a water molecule. Given a phospholipid molecule with fatty chains of 18 carbon atoms, it would be possible for the fatty chain of a phospholipid molecule to penetrate nearly to the middle of a membrane, while the polar head remained on the surface. All crevices could, therefore, be filled if the aliphatic chains penetrated the membrane from both sides.

The repeating units of the mitochondrial cristael membrane appear to be asymmetric with respect to the faces on which the polar heads of the phospholipid molecules are localized. If we assign a quasi-cuboidal geometry to the repeating unit, the top and bottom faces would be covered with the polar heads while the other faces would not. Such a distribution of the polar heads would require that the channels into which phospholipid molecules can slip, run in only one direction (between two of the faces and not between the other four faces).

BINDING BETWEEN REPEATING UNITS

When the major part of the phospholipid of a membrane is extracted with organic solvents, the repeating units, nearly phospholipid-free, may still remain attached to one another, and the original membrane structure can survive the extraction process. This result clearly demonstrates the facts that the protein of the repeating units constitutes the backbone of the membrane, and that phospholipid is not involved in the links between repeating units. These links involve primarily protein–protein interactions. However, phospholipid-free repeating units (after disaggregation of the membrane) no longer can reform membranes; only the phospholipid-containing repeating units still show this property. Phospholipid-free repeating units can combine with one another randomly to form a three-dimensional bulk phase (Figure III.7A).

MEMBRANE FORMATION FROM REPEATING UNITS

The negative charge of an array of phospholipid molecules on a surface of the cuboidal repeating units would prevent interactions with other repeating units at this surface. The interaction of such a negatively charged surface either with another negatively charged surface or with a hydrophobic surface would be interdicted (Figure III.7C). If we accept the thesis that all faces of the repeating units have combining capability, then only those surfaces of the repeating units which contain phospholipid, or those which are linked to the detachable sector, would be barred from interacting. If the phospholipid is localized on one surface of the cuboidal repeating unit and the detachable sector (as well as the phospholipid) on the opposite surface, such restricted interactions between repeating units would lead inexorably to the formation of a membrane (Figure III.7B and 7C). A corollary of this theory is that the capacity for membrane formation will be found in only a very limited class of protein units; the minimal requirement would be a cuboidal arrangement with only two of the opposite faces being specialized for binding to phospholipid or carrying a charge. In addition, the remaining two pairs of opposing

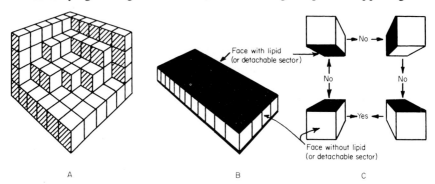

A   B   C

Fig. III.7. The restrictive role of phospholipid in the formation of membranes by repeating units. A: Represents the three-dimensional bulk phase formed by the repeating units which have been stripped of phospholipid. There are no restrictions of combining modalities in this type of association. B: Represents the two-dimensional continuum formed by repeating units which have the full complement of phospholipid. C: Represents the permissible and prohibited combinations between repeating units. (The presence of a detachable sector, instead of lipid, on one face of the repeating units, would lead to the same kind of membrane formation.)

faces would have to possess some complementary affinities, so that a two-dimensional membrane assembly would be more favorable than the monodisperse state which would result from the covering of all six faces with a charged phospholipid layer. This rather restrictive set of requirements, no doubt, accounts for the failure of mixtures of phospholipids and proteins which are not native parts of membranes, to form synthetic membranes.

MEMBRANE FORMATION AS MOLECULARIZATION

A membrane may be conceived of as a consequence of a special form of polymerization in which the combinations between repeating units can take place in only one plane—i.e. as directed two-dimensional versus random three-dimensional polymerization (Figure III.7). The directed polymerization leads to molecularization of the repeating units in the sense that the arrays of repeating units are one molecule thick; by contrast the bulk phase is a random, close-packed, three-dimensional jumble of repeating units. In the molecularized membranous state, the enzymic activities of the repeating units are measurable because the membrane is fully exposed to the solvent, hence to the solute molecules. By contrast, in the bulk phase this exposure is minimal; consequently, activity cannot be estimated. The essentiality of lipid for the enzymic activity of lipid-depleted repeating units usually has to do with this molecularization phenomenon.

NATURE OF MEMBRANE LIPIDS

The chemical nature of the phospholipid molecule—i.e. whether it is phosphatidylcholine or phosphatidylethanolamine or whether the fatty acid residues are mono-, di-, or polyunsaturated—does not affect much, if at all, the capability of phospholipid to induce *de novo* membrane formation. Any natural phospholipid can serve in that capacity. This is not to say that there is no specificity in the composition of phospholipids in membranes. There is, indeed, great specificity of composition, but this specificity is relevant to other properties of membranes (permeability, stability, etc.), not to membrane formation.

Each membrane has its own characteristic spectrum of phospholipid molecules (cf. Table III.1); the fatty acid residues of these phospholipids have characteristic degrees of unsaturation. The mitochondrial phospholipids are mixtures of phosphatidylcholine, phosphatidylethanolamine, and cardiolipin. No other phospholipid is found in any significant amount. There is relatively little neutral lipid in mitochondrial lipid, some 95 % of the total being phospholipid. The fatty acids of mitochondrial phospholipid are generally highly unsaturated—a fact that accounts for the great susceptibility of the isolated mitochondrial system to lipid peroxidation. Cholesterol is a major component of some membranes (e.g. the plasma membrane of the red blood corpuscle), but it is a minor component of the lipid

Table III.1

Phospholipid Composition of Several Membrane Systems

| Membrane system | Cells | Species | \multicolumn{10}{c}{Percent of total phospholipid[a]} |

Membrane system	Cells	Species	PC	PE	PS	PI	CD	SM	PG	PGP	GPG
Plasma membrane[b]	Red cell	Man	36	29	10	—	—	21	—	—	—
Plasma membrane[c]	Red cell	Chicken	42	23	5	2	—	21	—	—	—
Mitochondria[d]	Liver	Rat	49	30	—	8	12	—	—	—	—
Mitochondria[d]	Kidney	Rat	41	30	1	14	9	—	—	—	—
Mitochondria[d]	Heart	Sheep	46	35	—	5	10	—	—	—	—
Mitochondria[d,e]	Heart	Beef	41	33	—	8	15	—	—	—	—
Microsomes[f]	Liver	Rat	61	15	2.0	7	2.1	—	2.0	—	—
Lysosomes[f]	Liver	Rat	60	20	2.8	7.6	4.0	—	1.7	1	—
Plasma membrane[g]	Bacterial	*Serratia marcescens*	—	91	1	—	—	—	3	—	—
Plasma membrane[h]	Bacterial	*Halobacterium cutirubrum*	2	—	—	1	6	—	6	73	—
Plasma membrane[i]	Bacterial	*Bacillus cereus*	—	39	—	—	—	—	28	12	—
Chloroplast[j]	Plant	Spinach	39	2	—	13	—	—	—	—	46

[a] PC = phosphatidylcholine (lecithin); PE = phosphatidylethanolamine; PS = phosphatidylserine; PI = phosphatidylinositol; CD = cardiolipin; SM = sphingomyelin; PG = phosphatidylglycerol; PGP = polyphosphatidylglycerol; GPG = glycerolphosphorylglycerol.

[b] Farquar, J. W., and Ahrens, E. H., *J. Clin. Invest.* **42**, 675 (1963).

[c] Kates, M., and James, A. T., *Biochim. Biophys. Acta* **50**, 477 (1961).

[d] Strickland, E. H., and Benson, A. A., *Arch. Biochem. Biophys.* **88**, 344 (1960).

[e] Fleischer, S., Klouwen, H., and Brierley, G., *J. Biol. Chem.* **236**, 2936 (1961).

[f] Schwarz, H. P., Driesbach, L., Polis, E., Polis, B. D., and Soffer, E., *Arch. Biochem. Biophys.* **111**, 422 (1965).

[g] Kates, M., Adams, G. A., and Martin, S. M., *Can. J. Biochem. Physiol.* **42**, 461 (1964).

[h] Sehgal, S. N., Kates, M., and Gibbons, N. E., *Can. J. Biochem. Physiol.* **40**, 69 (1962).

[i] Kates, M., Kushner, D. J., and James, A. T., *Can. J. Biochem. Physiol.* **40**, 83 (1962).

[j] Wintermans, J. F. G. M., *Biochim. Biophys. Acta* **44**, 49 (1960).

of the mitochondrial membranes, representing less than 5% of the total lipid.

Evidence of an indirect nature supports the thesis that lipid forms a continuous phase which extends from repeating unit to repeating unit in the inner mitochondrial membrane continuum. In the hydrophobic sector of this lipid phase are dissolved mobile molecules such as coenzyme Q. Cytochrome c (a protein of 12,500 molecular weight) is known to bind electrostatically to phospholipids. It is, therefore, plausible to assume that cytochrome c binds electrostatically to the charged phospholipid heads, which line the surface of the cristael membranes. It could then "roll" or move about the surface of the membrane quite easily, fulfilling its function as electron carrier. Coenzyme Q would move about within the lipid-containing channels in the repeating units.

NATURE OF MEMBRANE PROTEINS

The protein molecules of repeating units (this generalization applies to basepieces and detachable sectors) are of two general kinds: catalytic proteins, i.e. functional proteins which are directly concerned in the catalytic processes; and noncatalytic proteins which are not so concerned. This is not to say that the noncatalytic proteins fulfill no function in the membrane. On the contrary, there is accumulating evidence that the noncatalytic proteins play a key role in membrane biogenesis, are the principal determinants of the ultrastructural pattern of repeating units, and confer control features on the catalytic proteins with which they associate. In membranes, generally, there appears to be a 1:1 relationship between catalytic and noncatalytic protein. This relation does not signify that each molecule of catalytic protein is always attached to a molecule of noncatalytic protein, and that such molecular duos are invariably the subunits of the various sectors of the repeating units of a given membrane. It would appear from studies of Complex IV that in this case at least the catalytic proteins interact as a set with a corresponding set of noncatalytic protein molecules.

There is not one unique noncatalytic protein in the mitochondrial membrane, but, rather, a group of such proteins. The basepieces of the inner mitochondrial membrane have a different set of noncatalytic proteins than do the headpieces.* Clearly, noncatalytic proteins have a certain degree of specificity, but not the complete specificity of catalytic proteins.

Noncatalytic proteins are highly versatile in combining capabilities. Three of the combining modalities are of particular relevance to their role in membrane phenomena: (1) combination of monomer with monomer to form a water-insoluble polymeric aggregate; (2) combination of polymeric noncatalytic protein with phospholipid to form a water-insoluble lipoprotein complex; and (3) combination of monomeric noncatalytic protein with monomeric catalytic protein to form a water-soluble complex.

Noncatalytic proteins can exist in a variety of conformational states: helix, β-conformation, random coil, and mixtures of these three states. Thus, structural protein in its polymeric water-insoluble form is largely helical in conformation. When structural protein is oxidized with performic acid or is succinylated, it undergoes depolymerization to monomeric units which are water soluble and exist in the random coil conformation. Any reagent which induces the random coil conformation depolymerizes polymeric structural protein into water-soluble monomers. The domains of structural protein which are in the helical conformation have a pronounced hydrophobic character whereas the same domains, when in the random coil conformation, have a polar character. The mole fraction of the nonpolar amino acids in structural and core proteins is not unusually high. Thus, it is not composition but conformation which is the principal determinant of the hydrophobicity of structural and core proteins.

The transition of noncatalytic proteins from one conformational state into another shows some unexpected hysteresis effects. It is as though there were an activation barrier to the transition of noncatalytic proteins from one conformational state into another. For example, structural protein can be isolated in a form which is water-soluble in aqueous media (e.g. a medium low in salt concentration and of pH about 8). Yet, polymeric structural protein will not dissolve even in 8 M urea or in 5 M guanidine. It has first to be depolymerized

---

* The noncatalytic proteins of the detachable sectors of the repeating units of the inner mitochondrial membrane are known as structural proteins, and those of the basepieces of these repeating units are known as core proteins.

by a combination of alkali and detergents before it assumes the conformation which permits solubility in aqueous media (the alkali and detergent being removed by dialysis).

The noncatalytic proteins of the basepieces of the inner mitochondrial membrane (the so-called core proteins) have structural features and properties which distinguish them from the noncatalytic proteins (structural proteins) intrinsic to the detachable sectors of the inner mitochondrial membranes. In general, the core proteins are insoluble in aqueous media—in some cases even in presence of 8 M urea, 5 M guanidine, or dilute alkali—and are barely attacked by proteolytic enzymes. Chemical alteration of core protein, e.g., by succinylation or by exposure to hydrazine, can lead to a water-soluble species. All structural proteins, on the other hand, have the property of water solubility when in the appropriate conformational state. Yet, despite differences in physical properties, there are sufficient points of similarity to sustain the thesis that core proteins and structural proteins are subgroups of a class of noncatalytic proteins. The similarities are found in the amino acid composition, peptide maps after tryptic cleavage, capability for combining hydrophobically with phospholipid, and high capability for helix formation.

HYDROPHOBIC BONDING AND SELF-ASSEMBLY OF MEMBRANES

It cannot be an accident or coincidence that the bonding by which a membrane is put together from its component parts is almost entirely hydrophobic or electrostatic. So far as can be ascertained there are *no covalent links* between proteins in any of the membranes yet examined. If, as appears probable, membranes are self-assembled from sectors of repeating units, then noncovalent bonding would be the only kind of bonding compatible with self-assembly. Covalent bonding would inevitably require the intervention of an enzyme and, thus, introduce complications in the assembly process. Electrostatic and hydrophobic bonds are the principal types that hold repeating units together and hold the parts of repeating units together. As yet there is little evidence of hydrogen bonding in any of the colligative interactions. Hydrogen bonding is of great importance in determining helical and β structures in proteins, but appears not to play a significant role in interparticle or intersector interactions.

The bonding between repeating units, between stalk and headpiece, and between the proteins within each sector is both hydrophobic and electrostatic whereas the bonding between phospholipid and protein and between basepiece and stalk is almost entirely hydrophobic. The nature of the bonding can be inferred from the reagents and conditions required to rupture the bonds. Thus cholate alone does not rupture bonds between repeating units whereas the combination of cholate and salt is effective. The requirement of both cholate and salt suggests hydrophobic and electrostatic bonding.

FUNCTIONAL UNITS IN MEMBRANES

The morphological sectors of a repeating unit (basepiece, stalk, and headpiece) have their counterparts in functional entities. Each of the complexes of the electron transfer chain is the functional counterpart of a basepiece of the inner mitochondrial membrane. There has to be this correspondence because the functional and the morphological expressions are two sides of the same coin. The demonstration of this correspondence has sometimes lagged behind prediction because the former depends entirely on technical consider- ations, the latter upon an understanding of the principles underlying membrane structure.

PERMEABILITY OF MEMBRANES

To round out a chapter on membranes we must give thought to the problems raised by the membrane as a barrier to the movement of ions and molecules from the external medium into the spaces bounded by the membrane. The fit of repeating units one to another is close enough to interdict massive penetration of the membrane by solute molecules and ions, but the interdiction is never complete. Even when the fit is perfect, there are inevitably spaces and channels by which small molecules can ultimately wind their way randomly through the membrane. The realities of hydrophobic and electrostatic bonding set a practical limit to the complete exclusion of solute molecules. Small molecules like those of water, oxygen, and CO_2, and ions such as the proton and hydroxyl ion, probably can penetrate the tightest of membranes (between repeating units), whereas larger molecules such as nucleotides and polypeptides have a low probability of penetration. Lipid-soluble molecules may enter the membrane via the lipid phase. This possibility enhances greatly the permeative capabilities of lipid-soluble molecules. Highly charged molecules are less likely to penetrate the membrane than are noncharged molecules of comparable size because electrostatic repulsion is likely to interpose difficulties. An important point to stress is that there are no *permanent* holes in membranes; the channels are statistical and impermanent. A membrane is a plastic, fluid system, and the fit of associated subunits is continually undergoing change and adjustment. That means that there is no permanent pathway for the random movement of molecules through a membrane.

There are various ways in which molecules can penetrate a membrane: (1) by random diffusion; (2) by facilitated diffusion: and (3) by an energized process.* In the latter two modes, penetration is not so dependent upon what, in the first case, resembles a random walk through an everchanging maze. In the case of facilitated or energized movement across a membrane, the repeating units of the membrane themselves play a role. These repeating units are permanent specific entities, in contrast to the ephemeral channels between them; thus, facilitated diffusion and energized transport can be very rapid and efficient processes.

SELECTED REFERENCES

BOOKS

Green, D. E., and Goldberger, R. F., "Molecular Insights into the Living Process," Academic Press, New York, 1967: see chapters on membranes.
Structure and Function of Membranes, *Brit. Med. Bull.* **24**, No. 2 (1968): a symposium volume on membranes.

SPECIAL ARTICLES

Benson, B. A., *J. Am. Oil Chemists Soc.* **43**, 265 (1966): on the orientation of lipids in chloroplast and cell membranes.
Criddle, R. S., and Park, L., *Biochem. Biophys. Res. Commun.* **17**, 74 (1964): on the structural protein of chloroplasts.

* These different modalities are mentioned here merely to remind the reader of the complexity of the problem of movement of solute molecules and ions through a membrane—a problem of immense biological and medical interest, which will be dealt with more fully in Chapter VIII.

Criddle, R. S., Bock, R. M., Green, D. E., and Tisdale, H. D., *Biochemistry* **1**, 827 (1962): physical characteristics of proteins of the electron transfer system and interpretation of the structure of the mitochondrion.

Fernández-Morán, H., Oda, T., Blair, P. V., and Green, D. E., *J. Cell Biol.* **22**, 63 (1964): a macromolecular repeating unit of mitochondrial structure and function.

Green, D. E., and Fleischer, S., *Biochim. Biophys. Acta* **70**, 554 (1963): the role of lipids in mitochondrial electron transfer and oxidative phosphorylation.

Green, D. E., and MacLennan, D. H., *Bioscience* **19**, 213 (1969): on the structure and function of the mitochondrial cristael membrane.

Green, D. E., and Perdue, J. F., *Proc. Natl. Acad. Sci. U.S.* **55**, 1295 (1966): membranes as expressions of repeating units.

Green, D. E., and Tzagoloff, A., *J. Lipid Res.* **7**, 587 (1966): role of lipid in the structure and function of biological membranes.

Green, D. E., Allmann, D. W., Bachmann, E., Baum, H., Kopaczyk, K., Korman, E. F., Lipton, S. H., MacLennan, D. H., McConnell, D. G., Perdue, J. F., Rieske, J. S., and Tzagoloff, A., *Arch. Biochem. Biophys.* **119**, 312 (1967): formation of membranes by repeating units.

Green, D. E., Haard, N. F., Lenaz, G., and Silman, H. I., *Proc. Natl. Acad. Sci. U.S.* **60**, 277 (1968a): on the noncatalytic proteins of membrane systems.

Green, D. E., Lenaz, G., Haard, N. F., and Silman, H. I., *Proc. Natl. Acad. Sci. U.S.* **60**, 277 (1968b): on the noncatalytic proteins of membrane systems.

Kopaczyk, K., Asai, J., Allmann, D. W., Oda, T., and Green, D. E., *Arch. Biochem. Biophys.* **123**, 602 (1968): on the resolution of the repeating unit of the inner mitochondrial membrane.

Korn, E. D., *Science* **153**, 1491 (1966): on the structure of biological membranes.

Lenard, J., and Singer, S. J., *Proc. Natl. Acad. Sci. U.S.* **56**, 1828 (1966): on a model of membrane structure.

Lenaz, G., Haard, N. F., Lauwers, A., Allmann, D. W., and Green, D. E., *Arch. Biochem. Biophys.* **126**, 746 (1968b): on the isolation and purification of mitochondrial structural proteins.

Lenaz, G., Haard, N. F., Silman, H. I., and Green, D. E., *Arch. Biochem. Biophys.* **128**, 293 (1968a): physical characterization of the structural proteins of mitochondria.

Martinosi, A., *Biochim. Biophys. Acta* **35**, 385 (1967): on the ultrastructure of the sarcoplasmic reticulum.

Murphy, F. A., *Science* **163**, 409 (1969): the ultrastructure of the influenza virus.

Perutz, M. F., Muirhead, H., Cox, J. M., and Goaman, L. C. G., *Nature* **219**, 131 (1968): on aqueous channels in hemoglobin.

Rabin, E. R., Jenson, A. B., Phillips, C. A., and Melnick, J. L., *Exptl. Mol. Pathol.* **8**, 34 (1968): ultrastructural study of herpes simplex virus.

Richardson, S. A., Hultin, H. O., and Green, D. E., *Proc. Natl. Acad. Sci. U.S.* **50**, 1821 (1963): on the universality of structural proteins in membrane systems.

Robertson, J. D., *in* "Cellular Membranes and Development" (M. Locke, ed.), p. 1, Academic Press, New York, 1964: on the phospholipid bilayer model of membrane structure.

Silman, H. I., Rieske, J. S., Lipton, S. H., and Baum, H., *J. Biol. Chem.* **242**, 4867 (1967): on the core protein of Complex III.

Sjostrand, F. S., *in* "Regulatory Functions of Biological Membranes" (J. Jarnefelt, ed.), p. 1, Elsevier Publishing Co., Amsterdam, 1964: on a model of membrane structure.

Sreter, F. A., Ikemoto, N., and Gergely, J., *Excerpta Med. Found., Int. Congr. Ser.*, **147**, 289 (1966): electron microscopic study of the sarcoplasmic reticulum.

Wallach, D. F. H., and Zahler, P. H., *Proc. Natl. Acad. Sci. U.S.* **56**, 1552 (1966): on a model of membrane structure.

Woodward, D. O., and Munkres, K. D., *Proc. Natl. Acad. Sci. U.S.* **55**, 872 (1966); *ibid.* **55**, 1217 (1966): on the structural protein of *Neurospora crassa*.

KEY REFERENCES FOR CHAPTER III

Repeating units of membranes

Fernández-Morán *et al.* (1964), Green and MacLennan (1969), Green and Perdue (1966), Martinosi (1967), Murphy (1969), Rabin *et al.* (1968), Sreter *et al.* (1966)

Composition of repeating units

Fernández-Morán *et al.* (1964), Green and Fleischer (1963), Green and MacLennan (1969), Green and Tzagoloff (1966), Green *et al.* (1968b), Lenaz *et al.* (1968a)

Bimodal nature of phospholipid

Green and Goldberger (1967), Chapter 3

Binding between phospholipid and protein

Benson (1966), Green and Fleischer (1963), Green *et al.* (1967), Lenard and Singer (1966)

Binding between repeating units

Green and Fleischer (1963), Green *et al.* (1967)

Membrane formation from repeating units

Green *et al.* (1967)

Nature of membrane phospholipid

Benson (1966), Green and Fleischer (1963)

Structural proteins

Criddle and Park (1964), Criddle *et al.* (1962), Green *et al.* (1968a,b), Lenaz *et al.* (1968a,b), Richardson *et al.* (1963), Silman *et al.* (1967)

CHAPTER IV

ULTRASTRUCTURE OF THE MITOCHONDRION AND ENZYME LOCALIZATION

THE TWO MEMBRANE SYSTEMS

The mitochondrial organelle is made up of two separable membrane systems—the boundary membranes and the cristael membranes. The two parallel boundary membranes enclose the mitochondrion and separate it from the suspending medium (see Figure IV.1A and B). The outer of the two boundary membranes is continuous and serves as one of the barriers to the movement of ions and molecules into the mitochondrial interior. The inner of the boundary membranes encloses the tubular cristael membranes. Each crista is a tube which is closed on its interior end and open on the end which fits into the inner boundary membrane. The space within the crista, thus, is continuous with the space between the two boundary membranes.* The cristae are not attached directly one to another,† but each crista is attached to the inner boundary membrane. In that sense, the cristael membranes are not continuous.

The two boundary membranes parallel one another with high fidelity and in the undamaged mitochondrion the width of the space separating these two boundary membranes is remarkably constant.

---

* In the strict topological sense, the cristae and the inner boundary membrane constitute one continuous surface, a hybrid membrane. Failure to appreciate the possible hybrid nature of this continuous surface had led to much confusion in the literature.

† In some mitochondria (e.g. the mitochondria of canary heart muscle) the cristae are joined by anastomoses.

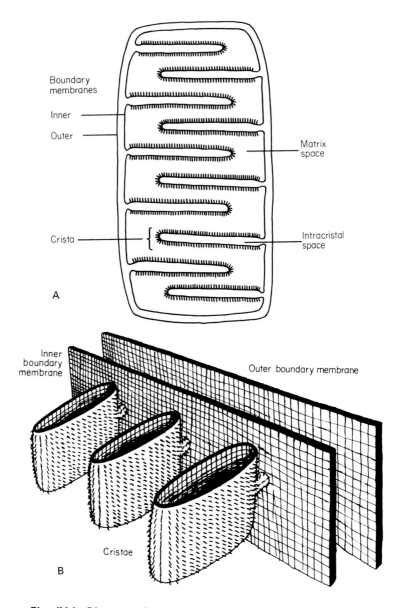

Fig. IV.1. Diagrammatic representation of the membranes of the mitochondrion. A: Whole mitochondrion. B: A three-dimensional view of a
section of the mitochondrion, showing particularly the relation of the cristae
to the inner boundary membrane and to the matrix space.

Under certain conditions of aging of the mitochondrion, the cristae pull away from the inner boundary membrane and form a compact mass within the interior of the mitochondrion or within a fold of the outer membrane (cf. Figure IV.2). The original points of attachment of cristae to inner boundary membrane are obliterated because of the tendency of open ends of membranes to reseal by rearrangement of repeating units. This sealing phenomenon accounts for the fact that the outer membranes can be isolated as a double-walled sac (this may be open at one end) with no attached cristae and with no remaining sign of the initial point of attachment of the cristae.

Electron micrographic studies of A. W. Linnane on the development of mitochondria in yeast exposed to antibiotics would suggest that mitochondria-like organelles* are formed which have the two boundary outer membranes but no cristae. If this interpretation is correct, then it would follow that outer and cristael mitochondrial membranes are assembled independently.

THE SPACES OF THE MITOCHONDRION

The mitochondrion, like other organelles, is a structured system with internal spaces which we may consider to be filled with aqueous fluid. Let us consider the subdivision of the mitochondrion with respect to these spaces. There is a space within the tubules of the inner membrane (the lumen of the cristae); a space between the two boundary membranes of the outer membrane system; and finally a space between the cristae (the matrix space). The lumen of the cristae in the orthodox mode is continuous with the space between the two boundary membranes, but, as we shall shortly see, this continuity is not always immediately obvious from the electron micrographs. We may treat the three spaces as separate and distinguishable with respect to the solute molecules contained therein even though there may be continuity between two of the spaces. The outer membrane system is more penetrable by small molecules than are the cristae. Thus, sucrose readily penetrates the outer membranes, but not the cristae. The sucrose-permeable space would then correspond to both the

---

* These organelles are essentially lacking in respiratory function and in tripartite repeating units.

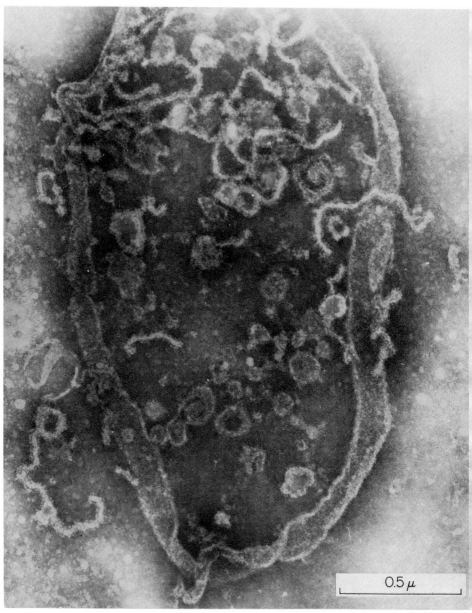

0.5 μ

Fig. IV.2. The separation of the cristae from the inner boundary membranes in aged mitochondria.

space between the two boundary membranes and the space between the crista in the orthodox mode. Although there is continuity of the lumen of the cristae in the aggregated and orthodox mode with the space between the two boundary membranes, the orifice is narrow so that the space is effectively impermeable to sucrose.

Water penetrates readily all spaces of the mitochondrion, and this is also true of small molecules such as ethylene glycol and glycerol. However, sugars or polyols with 6 carbon atoms or more do not readily penetrate the inner membrane, though they do penetrate the outer membrane. Ion pairs such as Na^+ and Cl^- or NH_4^+ and Cl^- readily penetrate both membranes, but other ion pairs such as Na^+ and acetate penetrate more slowly. A distinction has to be made between penetration of anions such as acetate *into* the structured elements of a membrane (by way of the lipid phase) and penetration of anions from one compartment to another *through* a membrane (by way of the spaces between repeating units). Acetate, but not chloride, can readily penetrate *into* the cristael membrane; chloride but not acetate can readily pass *through* the cristael membrane.

THE REPEATING UNITS OF THE MITOCHONDRIAL MEMBRANES

The cristae of the inner membrane have characteristic macro-tripartite repeating units, the individual parts of which are readily visualized in negatively stained specimens (see Figure III.3 for the repeat structures in the mitochondrial membrane and in the sarco-plasmic reticulum). The repeating units of the outer boundary membrane are not tripartite in structure.

DISTRIBUTION OF FUNCTION BETWEEN THE MEMBRANES

Appropriate chemical and physical methods have been devised for separating the boundary from the cristael membranes and for dis-secting the repeating units of the cristael membranes into their com-ponent sectors. By virtue of this separation and dissection it has been possible to pinpoint fairly accurately in which membranes, and in which sectors, the mitochondrial enzymic functions are localized. The distribution of function between the two membranes is neither a casual nor a random one. The inner membrane is the "combustion chamber" and contains the entire apparatus for generating ATP and for implementing the work performances powered by the energized state. The boundary membranes contain the enzymes of the citric cycle, the enzymes which carry out fatty acid oxidation and elongation, and various auxiliary enzymes such as the transaminases and transferases. These enzymes exist as ordered structures contained within the lumen separating the two boundary membranes and extending into the lumen of the cristael membrane. The ordered structure is stabilized by electrostatic interaction with the basepieces lining the lumen.

FUNCTIONS OF THE INNER MEMBRANE BASEPIECES

The monomeric species of the four complexes of the electron transfer chain have been identified in size, in shape, and in essential properties (e.g. the capability for *de novo* membrane formation) with basepieces of the inner membrane.* All the available evidence points to the basepieces as the locale of the electron transfer chain and of the system for generation of the energized state. Although a distinction between these two systems is often made, in reality none exists. Just as the armature in a dynamo generates electricity when rotated in a magnetic field, so the electron transfer chain of the intact mitochondrion, or of competent submitochondrial particles, inexorably couples electron flow to the generation of the energized state. However, it has proved impossible thus far to demonstrate in the isolated complexes of the chain the coupling of electron flow to the generation of the energized state. The coupling capability is *apparently* lost or destroyed when the reagents required to separate complex from complex are applied. Whether this loss is due to some structural rearrangement, or the exposure of a hitherto protected coupling site to the aqueous solvent, is not known. If the last were the case, then the coupling capability might still be retained, but might not be demonstrable with the presently available assay procedures owing to the rapid breakdown of the energized state. Electron transfer in the complex is probably always accompanied by conformational changes. In the isolated complex these changes may be uncontrolled

* There are other known species of basepieces besides the four complexes of the chain, and the possibility of yet other unrecognized species cannot be excluded.

as in an automobile in "neutral" with the wheels disengaged from the engine.

The inner membrane can be stripped of its detachable sectors; such stripped membranes can carry out coupled electron transfer, as evidenced by the catalysis of energized transhydrogenation. As far as can be ascertained, these stripped membranes contain only basepieces, i.e. only the complexes of the electron transfer chain.* From indirect evidence of this kind, it can be concluded that in the basepieces are localized the entire electron transfer chain (including the transhydrogenase complex), the capability for coupled electron transfer, and hence the capability for energized transhydrogenation.

FUNCTIONS OF THE INNER MEMBRANE HEADPIECE AND STALK

The headpiece and stalk play no part either in the electron transfer process or in the coupling of this process to the generation of the energized state, but they appear to be essential for the synthesis of ATP powered by the energized state, and for the generation of the energized state by ATP. The dependence of other performances, such as the translocation of divalent ions and energized swelling, on the headpiece and stalk sectors is still an unresolved point, though the available evidence supports this dependence. It is not possible at the moment to specify all the enzymes or systems which are localized in the headpiece and stalk. The isolated headpieces demonstrate an ATPase activity which is insensitive to the antibiotic oligomycin (or rutamycin) whereas the headpieces still linked to the stalk show an oligomycin-sensitive ATPase.† Thus, sensitivity to oligomycin (a reagent which specifically suppresses oxidative phosphorylation by blocking the terminal phase of the synthesis of ATP)‡ requires the attachment of the headpiece to the stalk.

* The complex concerned with transhydrogenation between TPNH and DPN⁺ is here considered as an electron transfer complex.

† The terminal phase of the synthesis of ATP is reversible so that the energized state can be generated at the expense of the hydrolysis of ATP. In the same way that electron transfer in isolated complexes can be "disengaged" from coupling, so may the terminal phase (in isolated headpieces or headpiece-stalks) be "disengaged." The result is the emergence of ATPase activities. In the case of the headpiece-stalk at least, such activity might actually represent the generation and spontaneous dissipation of a transient, energized conformational state.

‡ Oligomycin also blocks the generation of the energized state by ATP.

The headpiece is a composite of catalytic proteins and structural protein. The precise arrangement of the component proteins within the headpiece is unknown at present. What is known is that the mitochondrial oligomycin-insensitive ATPase can be purified extensively; and that such a purified ATPase corresponds in size and shape to a headpiece. It may be depolymerized into subunits, and this depolymerization results in loss of ATPase activity. Reassociation of these subunits into a headpiece-like form results in partial reconstitution of enzymic activity.

The stalk has been isolated by MacLennan and Asai and shown to be a single protein of about 25,000 in molecular weight. When the headpiece is separated from the stalk, the ATPase activity is insensitive to oligomycin whereas the ATPase activity of the headpiece-stalk unit is oligomycin sensitive. During the energy cycle the stalk undergoes a conformational transition from an extended to a relaxed state. The transition underlies the transfer of the energized state from the basepiece to the headpiece and makes possible the linking of the electron transfer process in the basepiece to the ATPase function in the headpiece.

THE ASSOCIATION OF ENZYMES WITH THE BOUNDARY MEMBRANES

There is a basic point of difference between the manner of association of the catalytic components of the complexes of the electron transfer chain with the basepieces of the cristael membrane and the form of association of the citric cycle enzymes with the boundary membranes. No simple way has been found for extracting the catalytic moiety of a complex of the electron transfer chain away from the basepiece of the inner membrane in which it is localized. The catalytic proteins are so tightly linked to the noncatalytic protein of the basepiece that separation of the two without complete loss of activity is impossible. Moreover, when separated the two moieties do not recombine to form the original basepiece. In contrast, the various enzymes of the citric cycle are readily detached from their association with the boundary membranes by relatively mild reagents and conditions. The residual, stripped membranes contain only noncatalytic proteins and associated lipid.

The enzymes localized in the boundary membranes are readily solubilized as soon as the outer boundary membrane becomes perforated or stretched. Since the links of the enzymic units with the boundary membranes are electrostatic, the presence of a monovalent salt in the suspending medium (0.01–0.1 M) is required for detachment and solubilization of these units. Beef heart mitochondria rapidly swell in distilled water and the outer boundary membrane becomes perforated toward the end of the swelling cycle. Nonetheless, the enzymes associated with the boundary membranes are not released until both detergent and salt are added to the suspending medium.

It has taken considerable time and much experimentation to appreciate the puzzling fact that a large number of enzymes are associated with, but are not intrinsic to, the boundary membranes. Both the inner and the outer boundary membranes can be completely stripped of all enzymic activities. Such stripped membranes probably contain exclusively or predominantly noncatalytic proteins. Smoly has shown that the enzymes associated with the boundary membranes form two-dimensional continua which are held together largely by hydrophobic bonds. On the basis of differences in extractability, it has been inferred that there are two different continua. It would appear that enzymes of relatively low molecular weight such as the isocitrate and malate dehydrogenases are in one of the two continua, while enzymes of relatively high molecular weight such as the pyruvate and α-ketoglutarate dehydrogenase complexes are in the other continuum. Each of the two continua is linked to the polar heads of the phospholipid molecules of the appropriate boundary membrane by electrostatic bonds. Once the continuum is separated from the boundary membrane it can readily break up into its component enzyme proteins. The continuum containing the enzymes of low molecular weight (the S fraction) is less tightly associated with its boundary membrane than the continuum containing the enzymes of high molecular weight (the K fraction). A clean separation of the enzymes of these two continua is, thus, readily accomplished. The S fraction contains all the enzymes of the citric cycle except for the pyruvate, α-ketoglutarate, and β-hydroxybutyrate dehydrogenase complexes. In addition it contains the phosphotransferases (nucleosidemonophosphokinase and -diphosphokinase). The K fraction contains the high molecular weight dehydrogenases specified above,

GTP succinyl thiokinase, and probably the enzymes which carry out fatty acid oxidation and elongation.

VECTORIAL CHARACTER OF THE BOUNDARY MEMBRANES

There is a wealth of evidence which bears on the polarity of the boundary membranes. There are catalytic sites on both the outer and inner surfaces of the outer boundary membrane. The same applies to the inner boundary membrane. Thus, clearly some of the enzymes of the outer membranes are accessible to the external medium. In addition, there is the entire group of readily solubilizable enzymes; these enzymes are separated from the external suspending medium by the outer boundary membrane. An enzyme system that converts fatty acids into the corresponding acyl-CoA esters behaves as if it were present definitely external to the outer boundary membrane; by contrast, the enzymes that oxidize the fatty acyl-CoA esters have been localized on the face of the boundary membrane that fronts the lumen. Since the outer boundary membrane is impermeable to acyl-CoA esters, it is not possible for acyl-CoA esters formed on the outside to undergo oxidation without further maneuver. This barrier to penetration is overcome by an interesting tactic (see Figure IV.3). By a displacement reaction, acyl-CoA esters are converted into acylcarnitine esters (carnitine replacing CoA). Acylcarnitine, thus formed, freely penetrates the outer boundary membrane. Once through the outer membrane, acylcarnitine is converted back into acyl-CoA by displacement of carnitine by CoA. The acyl-CoA ester then can be oxidized by the appropriate enzymes of the β-oxidation cycle. Even in isolated preparations containing fragments of the boundary membranes a requirement for carnitine in the oxidation of acyl-CoA esters is demonstrable. In the isolated membranes the original polarity with respect to localization of enzymes still remains. The enzymes which oxidize and transform acyl-CoA esters are localized in the interior space of the vesicular particles which are present in these crude preparations of the outer membrane.

The requirement for carnitine in fatty acid oxidation points up dramatically two features of the outer membranes: (1) the asymmetry of enzyme localization; and (2) the impermeability of the membranes to relatively large molecules such as CoA. The difference in size and in charge between carnitine and CoA is enough to tip the balance in favor of the entry of the carnitine derivative.

VECTORIAL CHARACTER OF THE INNER MEMBRANE

The repeating units of the cristael membrane have a built-in asymmetry. The headpiece-stalk sectors project from only one of the

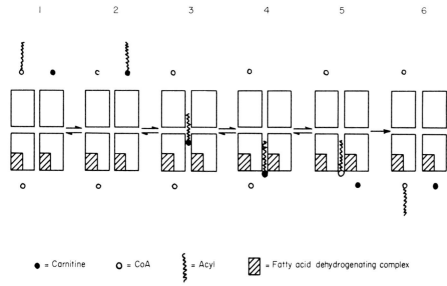

● = Carnitine   O = CoA   { = Acyl   ▨ = Fatty acid dehydrogenating complex

Fig. IV.3. A diagrammatic representation of the carnitine-mediated transfer of an acyl group from externally added acyl-CoA to internal CoA, and the requirement of carnitine for the oxidation of acyl-CoA by membrane fragments.

two faces of the basepiece which are parallel to the axis of the membrane. Let us designate that face of the basepiece as the H-S face and the parallel face on the opposite side of the membrane as the O face. In the mitochondrion, the solute molecules within the lumen of the crista can interact only with the O face, whereas in the submitochondrial particle, ETP_H, the solute molecules can interact only with the H-S face (see Figure IV.4). It is this reversal of directionality that accounts for two important consequences of the transition from

mitochondria to ETP_H. Cytochrome c can be extracted from mitochondria but not from ETP_H. Divalent metal ions can be translocated
by mitochondria but not by ETP_H. Cytochrome c is associated with
the O face of the basepiece. This face is accessible to the solvent in the
crista but not in ETP_H. Energized translocation of divalent ions is
initiated at the O face by interaction of the basepieces with divalent
ions in the solvent. The ions are translocated across the basepieces
to the space on the side of the H-S face. The cristae have the correct
polarity for this translocation but ETP_H does not. The solvent sees
the wrong face of ETP_H, hence no translocation is possible. It is

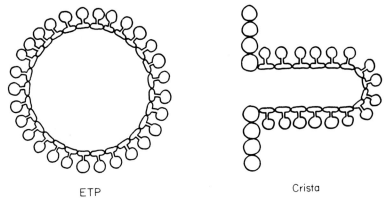

ETP   Crista

Fig. IV.4. A diagrammatic representation showing both ETP (with the
headpieces directed outward and the O faces of the basepieces inaccessible
to the solvent) and the crista (with the O faces accessible).

possible to prepare yet another submitochondrial particle (EP_1)
which forms vesicles like ETP_H, but the headpieces are directed to
the interior of these vesicles. EP_1 is able to carry out energized translocation just like the mitochondrion. Moreover, cytochrome c is
readily extracted from this particle.

ULTRASTRUCTURAL CHANGES AS AN EXPRESSION OF FUNCTION

Only a small beginning has been made in correlating the ultrastructural with the functional aspects of the mitochondrion. But the
considerable insight already gained in the rationalization of apparently
mysterious biochemical findings by reference to the ultrastructural

features provides a solid basis for expecting even more dramatic developments in the future.

We shall be dealing extensively with the ultrastructural changes relevant to the energized state in Chapters VI and VII. Our special concern in this chapter is with the rationale of the ultrastructural changes which accompany changes in the functional states of the mitochondrion. The form that a membrane assumes—i.e. whether a linear tubule or a zigzagging tubule, whether a vesicle or a tubule, whether a collapsed tubule or an expanded tubule—is determined almost exclusively by the geometry of the repeating units.* By geometry we refer not only to gross shape, but also to the distribution on the surface of the subunits, of charged and hydrophobic areas. If this geometry changes, then the form of the membrane will have to change in accommodation. Conversely, variability in membrane form means that there is variability in the geometry of the repeating units. If the change in geometry of repeating units can be correlated with changes in functional states, then the gross form of the membrane can be correlated with changes in functional states, since the gross form of the membrane is dictated by the geometry of the repeating units. In this way observed changes in morphology can be interpreted in functional terms.

The tripartite repeating unit of the inner membrane can exist in two extreme states (see Figure IV.5): (1) as the extended structure in which the stalk is fully extended, and basepiece and headpiece are maximally separated; (2) as the collapsed or condensed structure in which the stalk is contracted and the individual sectors of the repeating unit are no longer recognizable. The collapsed headpiece has the form of a disc $110 \times 110 \times 25$ Å. The detachable sectors are flattened on top of the membrane continuum during this transition. The type of membrane adjustment necessitated by conformational transitions in repeating units is beautifully illustrated in the spectacular electron micrographs taken by D. Slautterback of osmium-fixed sections of the mitochondria of canary heart *in situ*.† When the

---

* Other forces, such as osmotic pressure changes, might impose some deformation of the form determined by the geometry of the basepieces.

† No control was exercised over the state of the cristae in the canary heart mitochondria *in situ*. The different states were determined by intrinsic conditions within the heart muscle.

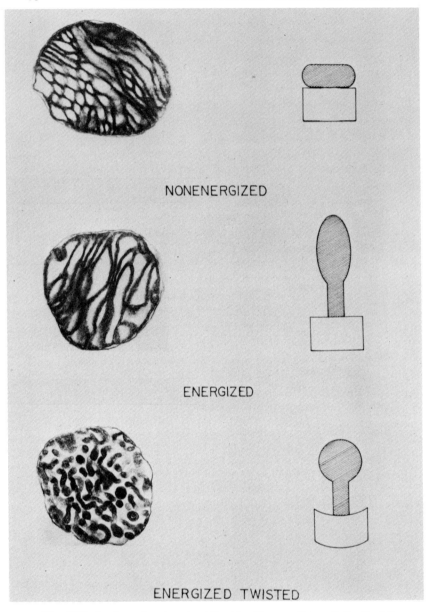

NONENERGIZED

ENERGIZED

ENERGIZED TWISTED

Fig. IV.5. Conformational cycle of the tripartite repeating unit. On the left-hand side of the figure are electron micrographs of isolated mitochondria in which the repeating units of the inner membrane are in the various conformational states illustrated by the diagrams on the right-hand side.

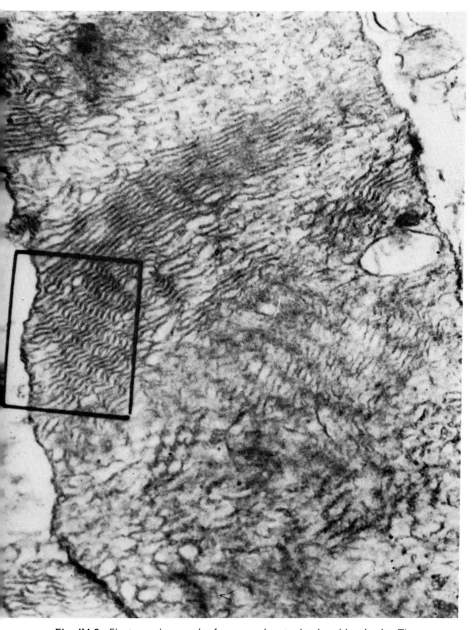

Fig. IV.6. Electron micrograph of a canary heart mitochondrion *in situ*. The micrograph was kindly provided by Dr. David Slautterback. The micrograph shows the different domains in which the cristael membranes are in one of several configurational states. The black frame encloses a domain in which the cristae are in the zigzag configuration—one of the energized states of the inner membrane. In other domains, the expanded vesicular configuration of the mitochondrion is clearly visible.

repeating units are collapsed, the cristae take the form of linear, flattened tubules (see Figure IV.6). When the repeating units are in the extended form, the cristae assume a periodic vesicular form which has the appearance of bubbles (see Figure IV.6). Another form which the cristael membrane assumes in these mitochondria is a remarkable zigzag configuration. As we shall be discussing in a later chapter, each of these configurations of the membrane can be interpreted in terms of particular conformational states of the repeating units.

When mitochondria are isolated in media containing sucrose (0.25 M or higher) the appearance of the cristae is vastly different from that of the cristae in mitochondria *in situ*. The presence of sucrose in the suspending medium compels an expansion of the crista to the point that neighboring cristae press one against the other. The apposition of the membranous walls of two expanded cristae leads to the aggregation of cristae and accounts for the change in geometry of cristae in isolated mitochondria. We shall refer to the state of the cristae in mitochondria *in situ* as the *orthodox* mode, and to the state of the crista in isolated mitochondria as the *aggregated* mode. The crista goes through exactly the same configurational cycle regardless whether it is in the orthodox or aggregated mode although on first inspection there may appear to be profound differences in geometry.

It is possible to modify experimentally the ultrastructural state of the cristae of isolated mitochondria. The endotoxin of *Bordetella pertussis* for example, can induce the cristae of isolated mitochondria to undergo a transition from the aggregated to the orthodox mode.

Isolated mitochondria, the cristae of which are in the aggregated mode, show three configurational states of the inner membrane—nonenergized, energized and energized-twisted (see Figure IV.5). Substrate or ATP induces the energized configuration; and substrate plus inorganic phosphate induces the energized-twisted configuration. The energized-twisted configuration of the crista in the aggregated mode is equivalent to the zigzag configuration of the crista in the orthodox mode (see Figure IV.6).

PERMEABILITY OF THE INTACT MITOCHONDRION

The mitochondrion under physiological conditions is largely an impenetrable unit. Small molecules like water, O_2, and CO_2 and also

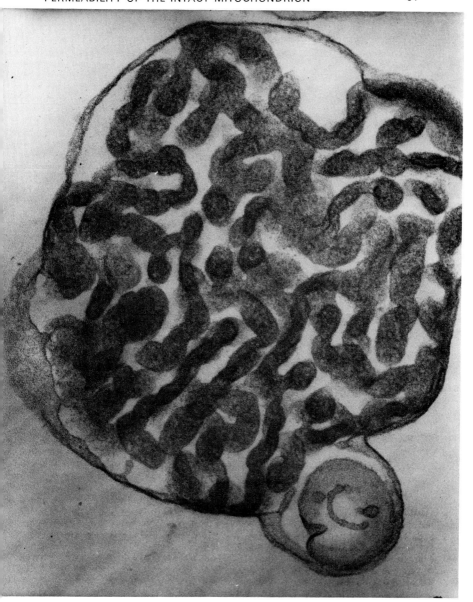

Fig. IV.7. Energized configuration of the inner membrane of isolated beef heart mitochondria in the presence of substrate and inorganic phosphate.

ions, such as protons, Na^+, Cl^-, acetate, do indeed penetrate the outer membranes rather rapidly. However (in the absence of special facilitating conditions), larger molecules such as citrate and oxalo-acetate, particularly when they have multiple charges, penetrate the outer membranes more slowly. Molecules which are even larger, such as ATP or DPN, do not penetrate at all (except possibly by facilitation under very special circumstances).

Thus, the mitochondrial membranes restrict severely the entry and exit of all but the smallest molecules and ions. This means that special devices have to be available for transmembrane transfer. ATP does not diffuse into and out of the mitochondrion. What is transported is a phosphoryl group, not ATP. The notion of the mitochondrion as a porous bag is without foundation when applied to an intact mito-chondrion immersed in a medium which preserves function. The impenetrability of the mitochondrion goes to the heart of a whole set of fascinating problems such as the need for carnitine,* the inhibition by atractyloside* of oxidation phosphorylation in intact mitochondria but not in submitochondrial particles, and the "opening phenomenon" (emergence of latent activities upon damage to the mitochondrial membrane).

The impermeability to relatively small molecules such as nucleotides makes it impossible for the mitochondrion to "lend" the electron transfer chain for direct oxidation of DPNH or TPNH generated external to the mitochondrion. This is not to say that all such collabor-ation between mitochondria and other organelles of the cell is excluded. There is accumulating evidence that various molecules can undergo a cycle of movement from outside to inside the mitochondrion and back again. During this cycle these molecules are metabolized first in one organelle and then in another—the mitochondrion being part of the circuit because it happens to contain enzymes necessary for certain key steps in an overall pathway. During other cycles, the cell makes use of an ingenious duplication of apparatus, namely specific dehydrogenases which are localized both inside the mitochondrion and in the extramitochondrial compartment. If the substrates for such dehydrogenases can penetrate the mitochondrion, then reducing equivalents generated extramitochondrially can be brought into the mitochondrion as reduced substrates.

* See Appendix II.

SELECTED REFERENCES

SPECIAL ARTICLES

Allmann, D. W., Bachmann, E., and Green, D. E., *Arch. Biochem. Biophys.* **115**, 165 (1966a): on the enzymes in the membrane-forming sectors of the outer membrane.

Allmann, D. W., Galzigna, L., McCaman, R., and Green, D. E., *Arch. Biochem. Biophys.* **117**, 413 (1966b): role of carnitine in oxidation of fatty acids by mitochondria.

Allmann, D. W., Harris, R. A., and Green, D. E., *Arch. Biochem. Biophys.* **120**, 693 (1967a): site of action of atractyloside; inhibition of fatty acid oxidation.

Allmann, D. W., Harris, R. A., and Green, D. E., *Arch. Biochem. Biophys.* **122**, 766 (1967b): site of action of atractyloside; inhibition of phosphoryl transfer.

Allmann, D. W., Bachmann, E., Orme-Johnson, N., and Green, D. E., *Arch. Biochem. Biophys.* **125**, 981 (1968): on the outer membranes of liver mitochondria.

Bachmann, E., Allmann, D. W., and Green, D. E., *Arch. Biochem. Biophys.* **115**, 153 (1966): on the enzymes of the spacer units in the outer membranes.

Bachmann, E., Lenaz, G., Perdue, J. F., Orme-Johnson, N., and Green, D. E., *Arch. Biochem. Biophys.* **121**, 73 (1967): the inner membrane of beef heart mitochondria.

Blondin, G. A., and Green, D. E., *Proc. Natl. Acad. Sci. U.S.* **58**, 612 (1967): on the factors which determine the penetration of ions into the mitochondrion.

Bremer, J., *J. Biol. Chem.* **237**, 2228 (1962): on the requirement of carnitine for fatty acid oxidation in mitochondria.

Byington, K. H., Smoly, J., Morey, A. V., and Green, D. E., *Arch. Biochem. Biophys.* **128**, 762 (1968): diethylstilbesterol solubilization of the outer mitochondrial membrane.

Chappell, J. B., and Crofts, A. R., *in* "Regulation of Metabolic Process in Mitochondria" (J. M. Tager, S. Papa, E. Quagliariello, and E. C. Slater, eds.), p. 293, Elsevier Publishing Co., New York, 1966: on the permeability of the mitochondrion to ions.

Fritz, I. B., *in* "Advances in Lipid Research" (R. Paoletti, and D. Kritchevsky, eds.), Vol. 1, p. 285, Academic Press, New York, 1963: on the role of carnitine in fatty acid oxidation.

Green, D. E., and MacLennan, D. H., *Biosciences* **19**, 212 (1969).

Green, D. E., and Tzagoloff, A., *Arch. Biochem. Biophys.* **116**, 293 (1966).

Green, D. E., Bachmann, E., Allmann, D. W., and Perdue, J. F., *Arch. Biochem. Biophys.* **115**, 172 (1966): on the enzymes in the outer membranes.

Green, D. E., Asai, J., Harris, R. A., and Penniston, J. T., *Arch. Biochem. Biophys.* **125**, 684 (1968).

Hackenbrock, C. R., *J. Cell Biol.* **30**, 269 (1966): on configurational changes in the membranes of liver mitochondria during controlled respiration.

Harris, R. A., Penniston, J. T., Asai, J., and Green, D. E. *Proc. Natl. Acad. Sci. U.S.* **59**, 830 (1968).

Kagawa, Y., and Racker, E., *J. Biol. Chem.* **241**, 2475 (1966): on the identification of oligomycin-insensitive ATPase with the headpiece of the tripartite repeating unit.

Kopaczyk, K., Allmann, D. W., Asai, J., Oda, T., and Green, D. E., *Arch. Biochem. Biophys.* **123**, 602 (1968): on the isolation and properties of an oligomycin-sensitive ATPase.

Lenaz, G., Littaru, G. P., and Castelli, A., *FEBS Letters* **2**, 198 (1969): on asynchronous development of different functions during mitochondrial biogenesis in yeast.

Lenaz, G., and MacLennan, D. H., *J. Biol. Chem.* **241**, 5260 (1966): on the non-extractability of cytochrome c from a submitochondrial particle (ETP_H).

MacLennan, D. H., and Asai, J., *Biochem. Biophys. Res. Commun.* **33**, 441 (1968): on the identification of the stalk with the protein which determines oligomycin sensitivity.

MacLennan, D. H., and Tzagoloff, A., *Biochemistry* **7**, 1603 (1968): on a low molecular weight protein which is the determinant of the oligomycin-sensitivity of mitochondrial ATPase.

O'Brien, R. L., and Brierley, G., *J. Biol. Chem.* **240**, 4527 (1965): on the permeability characteristics of isolated beef heart mitochondria.

Penniston, J. T., Harris, R. A., Asai, J., and Green, D. E., *Proc. Natl. Acad. Sci. U.S.* **59**, 624 (1968): on the geometric parameters of conformational changes in mitochondria.

Slautterback, D. B., *J. Cell Biol.* **24**, 1 (1965): on domains of configurational change in the cristae of canary heart mitochondria.

Wallace, P. G., and Linnane, A. W., *Nature* **201**, 1191 (1964): on mitochondrial profiles during mitochondrial biogenesis in yeast.

Williams, C., and Harris, R. A., in preparation: on the identical polarity of repeating units in cristae and EP_1.

KEY REFERENCES FOR CHAPTER IV

Two membrane systems of the mitochondrion

Allmann *et al.* (1968), Byington *et al.* (1968), Lenaz and Castelli (1969), Wallace and Linnane (1964)

Spaces in the mitochondrion

Allmann *et al.* (1968)

Distribution of function among the mitochondrial membranes

Allmann *et al.* (1966, 1968), Bachman *et al.* (1966, 1967), Byington *et al.* (1968)

Localization of function among the sectors of the tripartite repeating unit

MacLennan and Asai (1968), Green and Tzagoloff (1966), Green and MacLennan (1969)

Vectorial character of the boundary membranes

Allmann *et al.* (1966b, 1967a), Bremer (1962), Fritz (1963)

Ultrastructural changes in relation to function

Green *et al.* (1968), Hackenbrock (1966), Harris *et al.* (1968), Penniston *et al.* (1968)

Permeability of mitochondria

Blondin and Green (1967), Chappell and Crofts (1966)

CHAPTER V

MOLECULAR ASPECTS
OF THE MITOCHONDRIAL
SYSTEMS

One must not forget that the ultrastructural picture of the mito-chondrion as deduced from electron micrographs is a static one, as all morphological representations must be. But, the mitochondrion is a machine—pulsating and vibrating as the job is being done. It is to these dynamic aspects of the mitochondrion that the present chapter is addressed. To appreciate these aspects and to deal with them in the proper terms, we must concern ourselves with events and details at the molecular level.

MITOCHONDRIAL DYNAMICS

Imagine that we were to follow a cycle of mitochondrial activity from the time that a molecule of pyruvate collides with the appropriate site in one of the boundary membranes, to the completion of the cycle when ATP is formed and "delivered" to the outside of the mitochondrion. The pyruvic dehydrogenase complex is associated with the boundary membranes. It consists of a set of four enzymes and five cofactors which catalyze the following transformations: decarboxylation of pyruvate with formation of acetaldehydo-diphosphothiamine*; oxidation of acetaldehydo-diphosphothiamine with formation of acetyl lipoate* (the bound lipoate being reduced in the process); transfer of the acetyl group from reduced lipoate to CoA*; and finally the flavin*-mediated oxidation of reduced lipoate

---

* Diphosphothiamine, lipoate, and flavin are the tightly bound cofactors of the complex; CoA and DPN+ are the dissociable cofactors.

to oxidized lipoate by DPN^+. A unit of the dehydrogenating complex contains one molecule of each of the four enzymes. Once pyruvate interacts with the decarboxylase it is, so to speak, sucked into the membrane and directed along internal molecular tracks so that none of its carbon, hydrogen, or oxygen atoms leaves the mitochondrion except as CO_2 and water.* The conversion of pyruvate to acetyl-CoA takes place along the molecular tracks within the complex. In turn, acetyl-CoA progresses to the next enzyme in the sequence (citrate synthetase) without leaving the structured system of enzymes which is localized between the two boundary membranes. Exactly how the product of one enzymic complex is transferred to the next enzyme or complex in the sequence is still unknown. But, the probability is high that random molecular movements are minimized and that there are forces and devices which compel an orderly and directed progression of the constantly changing substrate from enzyme to enzyme and from complex to complex.

At the end of the citric cycle of events (after the expulsion of the end product, CO_2), all that is effectively left of a molecule of pyruvate is four pairs of electrons, which have been delivered to DPN^+ for the next round of activity. This round takes place in the electron transfer chain. We have ignored for the moment the fact that one of the intermediates of the citric cycle, succinate, has to move from the boundary membranes to the repeating units of the inner membrane, to deliver a further pair of electrons to the electron transfer chain. The oxidation product, fumarate, has then to return to the boundary membranes before the next step in the cycle, hydration of fumarate to malate, can be initiated.

The movement of succinate and DPNH from outer membrane to inner membrane is probably not a random one. As mentioned in the previous chapter there are physical links between the two membranes and these links may provide the molecular channels for the orderly, directed movement of DPNH and succinate from membrane to membrane.

The lipid phase of the inner mitochondrial membranes is un-

* We are referring here to events taking place under conditions that favor coupled respiration; under other conditions such as those favoring gluconeogenesis from lactate, there is a two-way traffic of organic molecules into and out of the mitochondrion.

doubtedly the major channel for the "directed" movement of mobile molecules such as coenzyme Q and cytochrome c. It may also fulfill a similar role for DPN^+,* succinate, fumarate, and possibly molecular oxygen. It would be not the entire molecule of DPN^+ or cytochrome c that abuts into the lipid phase, but only selected portions (possibly the pyridine ring of DPN^+ and the positively charged surface of cytochrome c). In this context the bimodal character of phospholipid must be borne in mind, i.e. the presence in the same molecule of both hydrophobic and polar sectors. Coenzyme Q would be "dissolved" in the hydrocarbon sector of the lipid; cytochrome c in the polar sector†; and DPN^+ possibly in both sectors.

Although the role of the continuous phospholipid phase in the outer membrane has yet to be assessed, it is eminently probable that this phase provides an ideal molecular track along which substrate molecules can be delivered from one repeating unit to another. Thus, there may be tracks within the sectors of the repeating units (built into the organization of each multienzyme complex) and tracks between repeating units (supplied by the continuous phospholipid phase).

In the inner membrane, electron donors generated by reactions in the outer membrane "deliver" electrons to the appropriate complexes (to Complex I by DPNH and to Complex II by succinate). Each pair of "delivered" electrons is transferred from complex to complex via the mobile components and finally transferred from Complex IV to molecular oxygen.

As the electrons are transferred along the potential gradient through each of the complexes (I, III, IV), their passage is accompanied by a molecular convulsion within the basepieces which leads to the energized state. This entails a profound change in the geometry of the

* DPN^+ and DPNH have been shown to form a complex with the β-hydroxybutyrate dehydrogenase–phospholipid system. This observation suggests the possibility that DPN^+ and DPNH may be transported within the lumen of the boundary membranes and within the extension of the lumen into the cristae of the inner membrane, in the form of a complex with a specific transport protein (or lipoprotein).

† A molecule of cytochrome c (which is positively charged) could be surrounded by a "shell" of phospholipid molecules (the negatively charged polar sector facing inward). Such a complex could now have a hydrophobic outer surface and be soluble in the nonpolar sector of the lipid phase.

repeating units of the inner membrane, expressed as a corresponding change in the configuration of the membrane. The transfer of the conserved energy from basepieces to headpiece with simultaneous synthesis of ATP again involves a conformational change in the repeating units which returns the system to the nonenergized state. Finally, the phosphoryl group of ATP, formed in the inner membrane, has to be delivered to AMP (or ADP) on the outside of the mito-chondrion. This is done by a series of phosphoryl transfers—from the headpieces of the tripartite repeating units across two membranes (cristael and outer membranes) and across the lumen separating these two membranes. It is known that the phosphoryl transferring enzymes are atractylate-sensitive, and that these are part of the structured com-plex of enzymes which is localized in the space between outer and inner membranes. The impermeability of the inner membrane to adenine nucleotides compels this elaborate sequence of phosphoryl transfer. How the phosphoryl transfer enzymes in the structured complex in the intermembrane space achieve this transfer across membranes is not understood.

This is only a partial chronicle of the train of molecular events which follow the interaction of pyruvate with the mitochondrial outer membrane. There is nothing static about this train of events. Everything is on the move—the substrate, the enzymes, the repeating units, the electrons, the phosphoryl group, the conformation of the repeating units and, as a reflection of these events, the membranes.

OPERATIONAL UNITS

The mitochondrion is a composite of a set of interdigitating, functional systems spread out between two membranes. Each of these systems is a complete operational unit—complete not only in respect to the enzymes as catalysts, but also with respect to the coenzymes as co-factors. The mitochondrion has a complete set of cofactors, all present at optimal levels, and constrained to remain within the mitochondrial membranes and interior. These cofactors include DPN^+, TPN^+, co-enzyme A, and the nucleotides of adenine, guanine, and cytosine. In general, these cofactors are bound to specific sites at which they participate in the appropriate catalytic processes. Other cofactors, such as the mobile carriers in electron transfer, are constrained not at specific sites but within a specific phase, the lipid phase, performing a shuttle service between the reducing and oxidizing systems.

The operational systems of the mitochondrion vary in complexity from a single enzyme (e.g. malate or isocitrate dehydrogenase), to systems of enzymes (e.g. pyruvate or α-ketoglutarate dehydrogenase), and finally to systems of complexes (e.g. the electron transfer chain). The citric cycle is implemented by two simple dehydrogenases (malate and isocitrate), two complex dehydrogenases (pyruvate and α-ketoglutarate), two hydratases (fumarase and aconitase), one electron transfer complex (succinate–coenzyme Q reductase), one condensing enzyme (acetyl-CoA oxaloacetate carboligase), one kinase (GTP-succinate thiokinase), and one transphosphorylase (nucleoside diphosphokinase). There are, thus, ten different operational units which are component entities in the conversion of pyruvate to CO_2 and reducing equivalents, by way of the citric acid cycle. These different units are within the boundary membranes (all but the succinate–coenzyme Q reductase) and in the cristael membrane (succinate–coenzyme Q reductase).

The systems for β-oxidation of fatty acids and for elongation of fatty acids are also multienzymic; these systems are probably associated with the boundary membranes. The same kind of logistic problems apply to the systems concerned in fatty acid oxidation and elongation as apply to the enzymes of the citric acid cycle.

The operational unit most characteristic of the mitochondrion, and probably of membrane systems in general, is the multienzymic complex. The four complexes of the electron transfer chain in the inner membrane, and the pyruvate and the α-ketoglutarate dehydrogenase complexes associated with the boundary membranes, are the most extensively studied systems in this category. There are, undoubtedly, additional complexes in the mitochondrion (concerned with fatty acid oxidation, fatty acid elongation, ATP synthesis, transhydrogenation, etc.), but these have yet to be characterized systematically. We may define a complex as a macromolecule built up of multiple proteins; each protein may be either an enzyme or a non-catalytic protein. In many cases it may be difficult for technical reasons to study and to specify the individual enzymes. The pyruvate- and α-ketoglutarate-dehydrogenase complexes illustrate some of the difficulties. In both these systems, the product of the reaction of one component in the coupled dehydrogenation becomes the substrate for the next reaction. When the product is a derivative of a bound coenzyme it becomes difficult to assay for a particular enzymic activity. Aldehydodiphosphothiamine is the product of the first reaction, and the substrate of the second. Acyl lipoate is the product of the second reaction, and the substrate for the third. In a complex of enzymes, the distinction between substrate and coenzyme becomes blurred because of the intimate physical association and arrangement of the enzymes which follow one another in the sequence. As we shall shortly see, the integration of function in the electron transfer complexes is so elaborate that it is difficult to write even a partial reaction catalyzed by one of the component catalytic species.

The complex is not only a complete operational unit, but it corresponds to or is associated with a morphological unit, i.e. to one or another sector of a repeating unit. The geometry of the multienzymic macromolecule is critical because the macromolecule has to fit accurately within the space limitations of a repeating unit. In the case of the complexes which are basepieces of the inner membrane, there is an additional requirement. These complexes must be linked not only to phospholipid, but also to their corresponding detachable sectors. Thus, the complex has to fulfill several needs—to serve as an operational unit, and also to participate in the structure of the membrane continuum.

INTERACTIONS WITHIN COMPLEXES

An exceedingly important question concerns the mechanism by which successive chemical interactions take place within a composite macromolecule such as a complex. How are coenzyme-bound products transferred, along the "internal molecular tracks" to which we have previously referred, from one enzyme to the next? The concept of the "swinging arm" has been invoked for this transfer by R. S. Criddle and R. M. Bock (see Figure V.1). Diphosphothiamine, lipoate, and flavin are in each case the three bound cofactors of the set of enzymes that catalyze the dehydrogenation of pyruvate and α-ketoglutarate to acetyl-CoA and succinyl-CoA, respectively. Each cofactor is an acceptor for one enzyme and donor for another (acceptor either of some chemical group or of electrons). If the coenzymes are anchored at some flexible group in the molecule (capable of rotation), and the rest of the coenzyme can oscillate about the flexible group, the distances through which the coenzyme can fluctuate may be

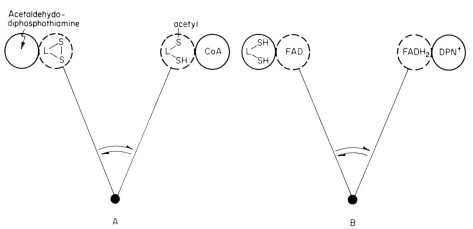

Fig. V.1. Diagrammatic representation of the swinging-arm hypothesis of Criddle and Bock. In A, the swinging group is lipoic acid; in B, FAD. Lipoic acid $\left(L\!\!<^{\;S}_{\;S} \right)$ is first reductively acetylated by acetaldehydodiphospho-thiamine, and in turn acetyl lipoate acetylates CoA. In the second sequence, FAD is reduced by lipoic acid $\left(L\!\!<^{\;SH}_{\;SH} \right)$ and in turn reduced FAD is oxidized by DPN^+. The swinging arm corresponds to a group of linked atoms such as a lysyl residue of the protein and the acyl chain of the lipoate moiety which is linked to it by a peptide bond. This group of linked atoms has free rotation about its fulcrum, i.e. the hindered atom in the protein to which the arm is attached.

sufficient to assure the making of molecular contact with each of the two enzymes it serves. Thus, the moving elements could be the bound functional groups of the complex.

A more sophisticated version of this swinging-arm hypothesis has been proposed by E. Frank Korman on the basis of atomic models. DPN^+ can be anchored to the protein by three means, through the carboxamide group of the pyridinium moiety, through the adenine ring, and finally through the pyrophosphate group.* In the reduction and oxidation of DPN^+ the pyridine ring undergoes a change in direction of orientation brought about by the transition from a tetrahedral to a trigonal N (see Figure V.2). This flip is of sufficient

* The latter two points of attachment being represented in Figure V.2 as R.

magnitude to enable the coenzyme to oscillate from interaction with a donor system to interaction with an acceptor system. This make-and-break feature of coenzyme molecules could apply to any transfer within a complex providing that the transfer changed the geometry of the coenzymic acceptor molecule. Thus, movement of coenzymes within a complex need not always be achieved by virtue of a flexible arm but may, in fact, be achieved by a conformational flip of the coenzyme during the transfer process.

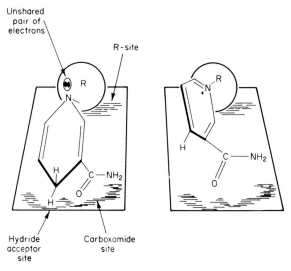

Fig. V.2. Conformational flip induced by the oxidation of enzyme-bound DPNH. It is assumed in this representation that DPN$^+$ is bound to the enzyme via both the carboxamide group of the pyridinium moiety and the pentose-phosphate-adenylate sector of the molecule (the R group).

The complexes of the electron transfer chain represent sets of oxidation-reduction proteins, Complexes I, III, and IV, each spanning a potential drop of sufficient energy to generate the energized state by a conformational rearrangement coupled to the transfer of a pair of electrons. Complex II also has a full set of oxidation-reduction proteins, but the potential drop within this complex is insufficient to generate the energized intermediate by conformational rearrangement.

The operational unit of electron transfer is the complex, not the individual cytochromes or other oxidation-reduction proteins in these several complexes. Two generations of biochemists have been indoctrinated with the notion that the components of the electron transfer chain are the individual cytochromes and flavoproteins. But the cytochromes in the chain (except for the mobile component, cytochrome c) are not the free monomeric species; rather they are subunits of multiprotein complexes. Electron transfer within a complex is not the same as electron transfer between monomeric species. The only process to which classical collision theory is applicable is the interaction of the mobile components with their respective donor and acceptor complexes. It was Keilin in the later 1920's who (in what was then a major conceptual advance) introduced the notion of individual oxidation-reduction proteins in the chain. Despite the accumulation of powerful evidence to the contrary, the notion of simple, single, sequentially interacting components has persisted for almost 40 years.

SELECTED REFERENCES

BOOKS

Green, D. E., and Goldberger, R. F., "Molecular Insights into the Living Process," Academic Press, New York, 1967.
Keilin, D., "The History of Cell Respiration and Cytochrome," Cambridge University Press, London and New York, 1966: a mine of information about the development of our present notions of the respiratory process.

SPECIAL ARTICLES

Allmann, D. W., Harris, R. A., and Green, D. E., *Arch. Biochem. Biophys.* **120**, 693 (1967a).
Allmann, D. W., Harris, R. A., and Green, D. E., *Arch. Biochem. Biophys.* **122**, 766 (1967b).
Green, D. E., and Allmann, D. W., *in* "Metabolic Pathways" (D. Greenberg, ed.), 3rd ed., Vol. 2, Academic Press, New York, 1968: chapter on fatty acid oxidation in mitochondria.
Green, D. E., and Fleischer, S., *Biochim. Biophys. Acta* **70**, 554 (1963): on the role of lipid in mitochondrial electron transfer and oxidative phosphorylation.
Green, D. E., and MacLennan, D. H., *in* "Metabolic Pathways" (D. Greenberg, ed.), 3rd ed., Vol. 1, Academic Press, New York, 1968: chapter on the mitochondrial system of enzymes.
Green, D. E., and MacLennan, D. H., *Biosciences* **19**, 212 (1969).

Green, D. E., and Tzagoloff, A., *Arch. Biochem. Biophys.* **116**, 293 (1966a): on the structure of the mitochondrial electron transfer chain.
Green, D. E., and Tzagoloff, A., *J. Lipid Res.* **7**, 587 (1966b): on the role of lipids in the structure and function of biological membranes.
Koike, M., Reed, L. J., and Carroll, W. R., *J. Biol. Chem.* **238**, 30 (1963): on the pyruvate and α-ketoglutarate dehydrogenating complexes.
Krampitz, L. O., Suzuki, I., and Gruell, G., *Brookhaven Symp. Biol.* **15**, 282 (1962): on the enzymic interaction of ketoacids with diphosphothiamine.
Massey, V., *Biochim. Biophys. Acta* **38**, 447 (1960): on lipoyl dehydrogenase.
Reed, L., and Cox, D. J., *Ann. Rev. Biochem.* **35**, 57 (1966): on lipoic acid.
Widmer, C., and Crane, F. L., *Biochim. Biophys. Acta* **27**, 203 (1958): on a lipid-soluble form of cytochrome *c*.

KEY REFERENCES FOR CHAPTER V

Enzymes of the citric acid cycle

Green and Goldberger (1967), Chapter 7, Green and MacLennan (1968)

Pyruvate and α-ketoglutarate dehydrogenase complexes

Green and MacLennan (1968), Koike *et al.* (1963), Krampitz *et al.* (1962), Massey (1960), Reed and Cox (1966)

Mobile molecules of the electron transfer chain

Green and Fleischer (1963), Green and MacLennan (1968), Green and Tzagoloff (1966b)

Complexes of the electron transfer chain

Green and MacLennan (1968), Green and Tzagoloff (1966a)

Transmembrane phosphoryl transfer

Allmann *et al.* (1967a,b)

Mechanism of electron transfer

Green and MacLennan (1969)

CHAPTER VI

THE
ENERGIZED
STATE

The major work processes in all cells are energized by ATP, or by surrogate substances, such as GTP or phosphocreatine, whose synthesis is mediated by ATP. The mitochondrion is charged with the task of generating the greater part of the ATP formed in all aerobic cells; it is this universal assignment which justifies the description of the mitochondrion as the power house of the cell. As we have discussed earlier, the electron transfer process is coupled only indirectly to the synthesis of ATP from ADP and P_i but is coupled directly to the generation of an energized state from which ATP is derivable by a subsequent interaction involving ADP and P_i. In the present chapter we shall be concerned neither with ATP formation nor with the generation of the energized state by ATP, but with the molecular strategy of coupling the electron transfer process to the generation of the energized state.

COMPLEXES OF THE ELECTRON TRANSFER CHAIN AS THE UNITS OF COUPLING

COUPLING AND ELECTRON TRANSFER COMPLEXES

The first of the two fundamental transductions of the mitochondrion is the reversible transformation of the utilizable energy released by an oxidative-reductive reaction into conformational energy. (The discussion of the second transduction, the conversion of conformational energy into the bond energy of ATP, will be taken up in the following chapter.) The free energy released in the oxidation-reductions of the

77

electron transfer chain is conserved in the conformational change. Such a conservation requires a mechanism for the direct transfer of the utilizable energy from the system undergoing oxidation to the system undergoing conformational change. The complexes of the electron transfer chain contain both systems. That is to say, the entire apparatus for coupling electron flow to the generation of the high energy state is contained within the precincts of each of the several complexes. Thus, coupling is an intrinsic part of the electron transfer process in the native complex. Damage of the complex can, of course, derange or modify the coupling capability, but the undamaged complex is the complete operational unit for coupled electron flow. In respect to being complete units, the complexes of the chain are analogous to dehydrogenase complexes both for pyruvate and α-ketoglutarate in which oxidation of aldehydodiphosphothiamine by lipoate is inextricably coupled to thioester bond formation (cf. Chapter VII).

Each of the complexes of the chain catalyzes an overall electron transfer sequence made up of the sum of some six component oxido-reductions. The drop in potential accompanying any one component oxidoreduction is insufficient for generation of the energized state. This is where strict analogy with the dehydrogenase complexes of the tricarboxylic acid cycle ends. For example, the potential drop between aldehydodiphosphothiamine and lipoate is sufficient for the genera-tion of a high energy bond. If the electron transfer process in the complexes of the chain involved a classical collision mechanism (i.e. if the individual components of the overall oxidation-reduction had an independent functional existence, as was generally assumed for a long time), the energy game would be lost because there would be no way of storing the free energy released in such a system until the total amount became sufficient for driving an endergonic reaction. The capability for conservation of free energy has to be built into a special-ized mechanism. The molecular strategy of the complex centers around the tactic for conservation of utilizable energy. Only the complexes of the electron transfer chain have the capability for using this tactic because the tactic depends upon the spatial interrelation-ships of a system of proteins capable of catalyzing oxidation-reductions, and not upon any single pair of the oxidative-reductive proteins within the complex.

We have mentioned in previous chapters that electron transfer in Complex II does not lead to the formation of ATP because the total drop in potential between the first electron donor for this complex (succinate) and the terminal electron acceptor (coenzyme Q) is insufficient to generate the energized state. Complex II is a "dummy" complex during the oxidation of succinate by coenzyme Q, but this is not to say that the complex lacks the capability for manifesting the energized state under all circumstances. There is indirect evidence that ATP can, in fact, generate the energized state in Complex II. In other words, the capability is latent and, although it is not manifested during electron flow from succinate to coenzyme Q, it can be induced by ATP. The enzyme complex which catalyzes either the oxidation of TPNH by DPN^+, or the oxidation of DPNH by TPN^+ shows similar behavior. The oxidation per se does not lead to the energized state because the drop in potential in the overall oxidoreduction is insufficiently large. However, ATP can induce the energized state of the reduced form of the transhydrogenating complex and thus can drive the reduction of TPN^+ by DPNH essentially to completion.

It would appear that all basepieces of the inner membrane, whether or not part of the electron transfer chain, are designed to manifest the energized state. As we shall discuss later, the energized state can be transmitted from basepiece to basepiece; given such transmission within a membrane, the notion of permanently incompetent complexes would be untenable.

THE TWO CONFORMATIONS OF COMPLEX III

The study of electron flow in Complex III has provided the clue to the strategy of energy conservation. This complex catalyzes the oxidation of reduced coenzyme Q by cytochrome c. In addition to its complement of noncatalytic protein (core protein), it is built up of two molecules of cytochrome b, one molecule of cytochrome c_1, one molecule of nonheme iron protein (containing two atoms of iron), and one molecule of a protein carrying the site which renders the complex characteristically sensitive to the fungicide antimycin. Electrons from external, reduced coenzyme Q are transferred directly to cytochrome b; then are transferred from reduced cytochrome b to cytochrome c_1; and finally are transferred from reduced cytochrome

c_1 to external cytochrome c. Since there are two molecules of cytochrome b and only one of cytochrome c_1, the transfer could either be stepwise (one electron at a time) or concerted (two electrons at a time), with some molecule other than cytochrome c_1 serving as electron acceptor. Complex III contains a nonheme iron protein which may well serve in that capacity. This protein undergoes oxidation-reduction during electron transfer in Complex III. The available evidence is consistent with nonheme iron as an additional link between reduced cytochrome b and cytochrome c. Thus, it is possible that one of the electrons is transferred via the sequence, cytochrome $b \rightarrow$ cytochrome $c_1 \rightarrow$ cytochrome c, while the other is transferred via the sequence, cytochrome $b \rightarrow$ nonheme iron protein $\rightarrow$ cytochrome c.

Antimycin prevents the passage of electrons from cytochrome b to cytochrome c_1. When the complex is inhibited by antimycin, it is impossible to separate cytochrome b from cytochrome c_1 by means, for example, of taurocholate plus ammonium sulfate; in absence of antimycin, this cleavage takes place readily. Reduction of the complex with dithionite has the same ultimate effect as antimycin; the reduced complex is no longer resolvable into cytochromes b and c_1 by taurocholate and salt. These fragmentation studies, as well as several other lines of evidence (such as changes in the number of exposed sulfhydryl groups) suggest that there are two states of the complex: an "open" state which is readily susceptible to fragmentation, and a "closed" state which is not susceptible. Antimycin and dithionite, respectively, freeze the complex in the closed state. In the absence of antimycin, the open state may be identified with the oxidized form of the complex; the closed state with the reduced form. For electron transfer to occur from cytochrome b to cytochrome c_1 a transition from the open to the closed form is requisite. Antimycin freezes the oxidized complex in the closed form, not by a reductive process but by a molecular rearrangement that leads to a state comparable to, but not identical with, the closed state that is achieved by reduction of the complex.

THE CONFORMATIONAL CYCLE OF THE TRIPARTITE REPEATING UNIT

Electron micrographic evidence points to a conformational cycle of the repeating units during coupled reactions. As shown in Figure

IV.5 the headpiece-stalk unit is collapsed atop the basepiece in the nonenergized conformation, and extended in both the energized or energized-twisted conformation. Thus, the headpiece-stalk unit undergoes a kind of jack-in-the-box movement during the conformational cycle. The electron transfer process is accompanied by a conformational change which leads to the separation of the headpiece-stalk unit from the basepiece. Only after the discharge of the energized state can the headpiece-stalk unit return to its original collapsed state.

We may equate the open state of Complex III with the nonenergized conformation of the repeating unit, and the closed state of Complex III with the energized conformation of the repeating unit. The electron transfer process cannot proceed when the repeating unit is in the energized conformation. What accounts for this block induced by the conformational change? The phenomenon could be accounted for in terms of the following hypothesis. The oxidized form of cytochrome b in the energized state of the complex is assumed to be nonreducible by reduced coenzyme Q. Let us say that the potential of cytochrome b becomes too negative to permit reduction by reduced coenzyme Q. Thus, no further electron transfer can take place until the complex becomes deenergized. Then, in the deenergized complex cytochrome b becomes reducible by reduced coenzyme Q, and electron transfer can resume once again.

The events described above for Complex III appear to apply with equal force to the other complexes involved in coupled reactions (see Figure VI.1). The oxidoreductions within the complex lead to the conformational change. The interactions between the mobile reductant or oxidant and the complex are concerned with the regeneration of the proper oxidation-reduction state of the complex and not with the conformational changes involved in energy conservation.

ENERGY CONSERVATION DURING DISCONTINUOUS ELECTRON FLOW

The working hypothesis is that the conformational changes described above as accompanying electron transfer in Complexes III and IV, and presumably in Complex I as well, are of sufficient magnitude to drive the eventual formation of ATP. The utilizable energy released by the oxidoreductive process is translated into conformational and electrostatic energy and stored in these forms.

The conformational change of the complete cycle conceivably could represent a free energy equal in magnitude to the drop in potential, i.e. to the sum of the free energy released by one electron falling between reduced coenzyme Q and cytochrome c_1 plus the

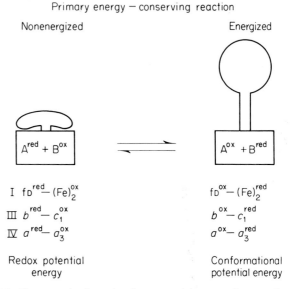

Fig. VI.1. The transduction of redox potential energy into conformational potential energy in the transition of the tripartite repeating unit from the non-energized to the energized conformation. In this formulation it has been assumed that the electron transfer components of each of the three complexes of the chain which are involved in coupling (Complexes I, III, and IV) can be segregated into two groups—A and B. In the nonenergized conformation of the repeating unit, the components of the A group are reduced and those of the B group are oxidized. Electron transfer leads to the oxidation of the components of the A group and reduction of the components of the B group. Only one representative of each of the groups has been shown in the diagram. The symbol f_D represents the flavoprotein of Complex I; $(Fe)_2$ represents one of the nonheme iron groups of the same complex; b, c_1, a, and a_3 represent the various cytochromes. We are indebted to Dr. G. Vanderkooi for the design of this figure.

free energy released by a second electron falling between reduced coenzyme Q and nonheme iron. This would be fairly close to the maximal free energy released by the overall oxidation (oxidation of reduced coenzyme Q by cytochrome c). In theory, at any rate, the ultimate formation of ATP is possible by this mechanism, if the conversion of utilizable energy released by the oxidative reaction to conformational energy and the subsequent conversion of conformational energy to the bond energy of ATP are both highly efficient.

Maximal conservation of utilizable free energy released in a chemical reaction requires that the system operate at equilibrium.* Can this condition be fulfilled in a biological system which is constantly in cycle? There is, in theory, a simple way out of the dilemma of how to carry out the reaction at equilibrium. If the primary system in which ATP is generated achieves equilibrium extremely rapidly as compared to the subsequent, much slower process (delivery of the phosphoryl group to external ADP), then, indeed, essentially maximal efficiency of conservation can be achieved in the primary coupling system.

STRUCTURE OF COMPLEX IV

It has been possible to resolve Complex IV into a moiety containing catalytic proteins and another moiety containing core proteins—the cross-contamination between the two being small. Such a resolution would argue that in the native complex the catalytic proteins formed one combined set which was, in turn, linked to another set of combined core proteins. This interrelationship between catalytic and core proteins, which is probably true for all four complexes, means that conformational change in the complex cannot be localized exclusively in either the catalytic proteins or the core proteins since we are dealing with one integrated system. The core protein has to be visualized as an intrinsic part of the electron transfer chain, not as a protein extraneous to the chain. The core protein appears to be capable of extensive conformational change, and this specialized capability is probably highly relevant to the conservation of energy in the coupling process. The name "core protein" might prove to be unfortunate in that it implies that the noncatalytic protein is buried within the interior of a complex. The exact positional relationship between catalyic and

* Another consequence of the operation of the system at equilibrium, i.e. the reversal of the direction of electron flow in energized complexes, is considered in Chapter VII.

noncatalytic protein within a complex is at present unknown. Therefore, the designation "core protein" should not be taken literally.

ON THE NATURE OF THE ENERGIZED STATE

The classical view of the coupling of electron transfer to the synthesis
of ATP centered around the notion that the electron transfer process
is coupled to the synthesis of a primary high energy intermediate
(not ATP) which, by a series of substitution reactions, is eventually
converted into ATP. According to this view there is a direct conversion of the free energy liberated in the oxidative reactions into the
bond energy of some unspecified high energy intermediate. (The
theory would further require that the "first high energy intermediate"
is not a phosphoryl derivative.) The inability of two generations of
biochemists either to isolate or to demonstrate these postulated high
energy intermediates should have engendered grave suspicions as to
the reliability of the assumptions underlying the classical interpretation of coupling. Oddly enough, such an evaluation has rarely been
made. Rationalization of the failure to isolate intermediates was in
terms of technical difficulties attending such isolative attempts, not
in terms of inadequacy of the basic assumptions.

It was Peter Mitchell who more than anyone else persistently
suggested an interpretation of coupling in terms other than those of
high energy intermediates. Although the correlation of coupling with
proton movements* is probably not primary, nonetheless the theory
of Mitchell had the virtue of reminding biochemists that there are
other ways of interpreting the coupling of electron transfer to synthesis
of ATP and that failure to observe or to isolate high energy intermediates could simply mean that such intermediates do not exist.

---

* In the Mitchell "chemiosmotic hypothesis," electron transfer is visualized as
achieving a transfer of protons across the membrane. The "proton-motive force"
thus generated is held to represent the conservation of the energy of oxidation-
reduction. This "p.m.f." has the capability of driving in reverse a proton-trans-
locating, anisotropic ATPase, thus achieving oxidative phosphorylation. For a full
treatment of this ingenious hypothesis see *Biol. Rev. Cambridge Phil. Soc.* **41**, 445
(1966). Experimental evidence from several laboratories is now available which
points up the severe limitations of the chemiosmotic interpretation of energy
transduction.

The thesis which we are developing in the present volume is that the energy transduction involved in the mitochondrial coupling of electron transfer to synthesis of ATP in the inner membrane is the conversion of the free energy released by oxidoreduction to conformational and electrostatic energy. ATP synthesis requires yet a second transduction, namely the conversion of conformational plus electrostatic energy into the pyrophosphate bond energy of ATP. This interpretation rests in large measure on the experimental fact that concomitant with the coupling process, the inner membrane undergoes profound rearrangements in shape and pattern—rearrangements which can be traced to the conformational cycle of the repeating units described above.

It could be argued that the conformational changes are secondary events in coupling and that these changes are not necessarily relevant to the problem of energy transduction. There are five considerations that militate against such a watered-down interpretation of the conformational changes accompanying coupling. First, apart from the gross changes in membrane configuration, is the fact of discontinuous electron flow accompanied by changes in the conformational states of the complexes. Second is the sheer magnitude of the gross membrane changes. If these changes were unrelated to the primary transducing events then the efficiency of the transduction would have to be low since these changes would represent dissipation of free energy. The coupling efficiency of the mitochondrion is known to be high; and such high efficiency is incompatible with the notion that coupling involves a great deal of "waste motion" in the form of conformational change. The third consideration derives from the evidence that conformational change similar in nature to that observed in the mitochondrion is an invariant feature of energy transduction in all membranes tested. The universality of the correlation between energy transduction and conformational change in membranes again argues against a secondary role for conformational change in the coupling of electron transfer to synthesis of ATP. Fourth, mitochondria can be partially cross-linked by treatment with bifunctional reagents without loss of electron transfer activity. Such cross-linked mitochondria are restricted in the conformational changes they can undergo; they do not exhibit any energy-linked functions. Finally, the speed of conformational change is of the same order of magnitude as the turnover of the components of the electron transfer chain.

The electron microscopic evidence referred to above bears on the gross changes in the membrane and also on the changes in the geometry of repeating units during coupling. This evidence of configurational change can now be supplemented with independent evidence of extensive changes in light scattering during the transition of mitochondria from the nonenergized state to the energized. In fact methods in profusion are now available for demonstrating conformational changes in the mitochondrion during the energy cycle. Reagents which exhibit marked electron spin resonance have been introduced into the cristael membrane, and the change in spin resonance during the energizing process has been used as an index of conformational change. Similarly reagents which fluoresce have been used as indicators of change in the conformation of the repeating units during the energy cycle.

ENERGETICS OF CONFORMATIONAL CHANGE

The definition of the conformational cycle by direct electron microscopic observation has made it possible to recognize the gross events which underly conformational change and which are consequences of this change. The most obvious event is the disappearance of the stalk in the nonenergized conformation and its extension in the two energized conformations. The stalk either sinks fully extended into the basepiece or alternatively it collapses. A piston action of the stalk poses various logistic difficulties and it is more likely that the stalk is expanded in the energized conformation and relaxed in the nonenergized conformation. Such a transformation is readily accounted for by a transition from a helix to a random coil. We are postulating that conformational change in the basepiece accompanying electron transfer induces helix formation in the stalk and extension of the headpiece-stalk unit. Helix formation makes the stalk rigid and compels the transition from a collapsed to an extended headpiece. In this way the stalk acts as a transmitter of the energized state from basepiece to headpiece. The conformational perturbation is thus transferred from basepiece to headpiece. The rearrangement of the headpiece from a collapsed disc to an expanded sphere during the energizing of the repeating unit probably is more than a simple lifting of the headpiece away from the basepiece. The extension of the stalk could compel a major rearrangement of the proteins in the headpiece. When ATP is synthesized in the headpiece, sufficient conformational energy has to be available in the headpiece to drive the synthesis.

The headpiece-stalk units of cristae in the aggregated mode are directed exteriorily, i.e., into the space between cristae. The conformational cycle is bound to affect profoundly the magnitude of the Donnan effect. When the headpiece-stalk unit is collapsed atop the basepiece, there is extensive screening or neutralization of the charges of the polar heads of the phospholipid molecules which cover the surface of the basepiece. When the headpiece-stalk unit is extended, the Donnan effect is increased because a sizeable number of the charged groups of the phospholipids become uncovered. Conformational change, thus, inexorably modifies the magnitude of the Donnan effect. With this modification cations must enter or leave to achieve electrical neutrality; the entry or exit of cations means that protons must also participate in this charge readjustment. Conformational change leading as it does to change in the Donnan equilibrium inevitably leads to proton and cation movements. It is not surprising, therefore, that changes in pH parallel exactly the transition from the nonenergized to the energized conformation.

The fundamental question that remains to be resolved is how conformational energy can be transduced into the bond energy of a pyrophosphate in ATP. In principle one can specify how that could be accomplished. One possibility is that conformational change could lead to the formation of an electrostatic well in the headpiece in which a set of five positive charged groups are brought into close apposition. When ADP^{3-} and P_i^{2-} slip into this electrostatic well, and bind to the appropriate sites, all the electrostatic and entropic barriers to interaction are eliminated; the combination of ADP and P_i in this special environment would then be a reversible process involving a very small free energy change. It is as if an enzyme had to be brought to the active form by a preliminary endergonic conformational change, and once so energized the enzyme could catalyze the reversible combination of ADP and P_i. Conformational energy would be used to set up the special electrical and geometric environment in which an isoergonic combination of ADP and P_i can take place.

Another possibility is suggested by some ultrastructural observations. Engagement of cristae is an essential condition for coupling. This engagement which leads to the energized conformation of the repeating units involves an interaction between the headpieces and basepieces of apposed cristae. A low energy compound may thereby be formed. Inorganic phosphate ruptures the bond linking headpiece to basepiece, becomes covalently linked to the headpiece, and induces the energized-twisted conformation. A conformational rearrangement of the system induced by ADP could trigger the transition from a low to a high energy phosphate group.

PSEUDOENERGIZED SWELLING

Mitochondria suspended in a sucrose-containing medium can be induced to swell in presence of salts to which the inner membrane is permeable (e.g. KAc) providing the process is energized by electron transfer or by hydrolysis of ATP. It is also possible to induce mitochondrial swelling with appropriate salts (e.g. NaAc) in absence of an energy source providing the mitochondria are in sucrose-free media. Only salts which penetrate the inner membrane (both through the membrane and into the membrane) are competent as inducers of this type of swelling. No energy source is required for swelling induced by salt under these conditions. This type of swelling is referred to as pseudoenergized swelling since the salts can spontaneously induce the same kind of configurational changes in the cristae as are observed in energized swelling.

Later, we shall be considering in some detail the nature of the ultrastructural and ionic transitions involved in swelling. Our immediate concern is with the mechanism by which appropriate salts can induce the ultrastructural equivalent of the energized and energized-twisted configurations in absence of an energizing source. Moreover, a satisfactory explanation has to be found for the strange fact that the pseudoenergizing of the membrane is unaffected by uncouplers or inhibitors of electron transfer or of hydrolysis of ATP.

The conformational change induced by the electron transfer process can be duplicated by ion pairs such as NaAc or by a high internal pH. But, unlike the energized process, there is no cycling of the repeating units. The conformational change is frozen as it were when induced by NaAc or high pH. What this induction means is that under the new conditions (high internal concentration of NaAc or high internal pH) the free energy change in the transition from the "nonenergized" to the "energized" conformation is small or negative. The transition can then proceed spontaneously. The mechanism of induction of conformational change by electron transfer is different from the mechanism of induction by NaAc or high pH. Hence, reagents like uncouplers which affect the mechanism involving electron transfer have no effect on the mechanism involving salt or high pH.

MULTIPLE WORK PERFORMANCES

The energized or energized-twisted state can be discharged in a variety of ways. Interaction of the energized-twisted state with ADP leads to synthesis of ATP and discharge to the nonenergized state. Interaction of the energized-twisted state with Ca^{2+} leads to the deposition of $Ca_3(PO_4)_2$ and the discharge of the energized state. Interaction of the energized-twisted state with monovalent cations leads to the translocation of monovalent salts and the discharge of the energized state. In this exceptional instance anions other than phosphate can induce the energized-twisted state. It is probable that the energized state rather than the energized-twisted state is coupled to transhydrogenation. The system, reduced by DPNH, is energized by ATP and discharged by TPN^+. The important point at issue is that all the work performances involve the same basic coupling system. What determines whether this or that work performance will take place is the presence or absence of the discharging agent. If Ca^{2+} is present, translocation of Ca^{2+} takes place; if ADP is present, synthesis of ATP takes place. But if both Ca^{2+} and ADP are present, then the reagent with higher affinity for the site of discharge will win the day. If the affinities are equal, both work performances may proceed simultaneously. The only option may be energized-twisted versus energized state. Three of the work performances are geared to the energized-twisted state, and only one to the energized state.

TRANSMISSION OF THE ENERGIZED STATE

Particles can be prepared which do not oxidize pyruvate plus malate, or succinate, but retain unimpaired the capacity to oxidize ascorbate plus tetramethyl-*p*-phenylenediamine.* When such particles are examined by electron microscopy under conditions suitable for generation of the energized state, it is found that the entire inner membrane assumes the energized state. If only the complex capable

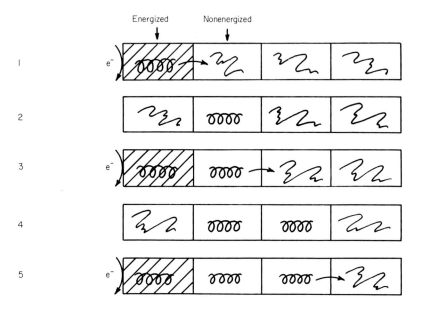

Fig. VI.2. Transmission of the energized state through the cristael membrane. The energized state (represented by the regular loops) is generated at one site (represented by cross hatched lines) and then transferred in any direction to another site. Each site corresponds to a single repeating unit of the membrane continuum. The transfer involves no loss of energy—only a change in locale of the energized state. The donor site becomes deenergized while the acceptor site becomes energized. This is a random exchange process which is presumed to be rapid compared to the rate of generation of the energized state in the membrane. The key notion implicit in this diagram is that there can be no localization of the energized state in a membrane. The energized state will radiate through the membrane as quickly as it is generated at any particular site and this will proceed until all the sites become energized.

* An artificial substrate system which feeds electrons into the chain at a potential close to that of reduced cytochrome *c*.

of electron transfer (Complex IV) became energized and the other complexes did not, the membrane would not be able to undergo a complete transition in form. In fact, it can be shown by many variations of the experiment cited above that electron transfer restricted to any one of the complexes is sufficient to energize the entire inner membrane. Moreover, this generalization holds true even when the possibility that ATP might serve as an intermediary energizing source is completely excluded by the presence of oligomycin (rutamycin). Further, oxidation of ascorbate plus tetramethyl-p-phenylenediamine by Complex IV can energize the reversal of electron flow through Complex I, even in the presence of rutamycin, which prevents ATP formation, and in the presence of antimycin A, which prevents electron flow through Complex III. These observations probably mean that the energized state generated at one site can radiate throughout the membrane by a sort of exchange reaction (see Figure VI.2). In other words there can, in effect, be a transfer of the energized state from one complex to its neighbor. The donor complex is no longer energized after the transfer; the acceptor complex, however, becomes energized. The physical mechanism by which the energized state is transmitted through a membrane from one repeating unit to its neighbor repeating unit is unknown. But in a general way transmission of the energized state through a membrane is an expression of the cooperative properties of membranes.

SELECTED REFERENCES

BOOKS

Green, D. E., and Goldberger, R. F., "Molecular Insights into the Living Process," Academic Press, New York, 1967: see Chapter IX on energy transductions.
Lehninger, A. L., "The Mitochondrion," W. A. Benjamin, Inc., New York, 1965.

SPECIAL ARTICLES

Baum, H., and Rieske, J. S., *Biochem. Biophys. Res. Commun.* **24**, 1 (1966): on proton involvement in electron transfer.
Baum, H., Rieske, J. S., Silman, H. I., and Lipton, S. H., *Proc. Natl. Acad. Sci. U.S.* **57**, 798 (1967); *J. Biol. Chem.* **242**, 4876 (1967): on the mechanism of electron transfer in Complex III.
Blondin, G., Vail, W. J., and Green, D. E., *Proc. Natl. Acad. Sci. U.S.* **58**, 612 (1967); *Arch. Biochem. Biophys.* **129**, 158 (1969): on pseudoenergized swelling.

Boyer, P. D., *in* "Oxidases and Related Redox Systems" (T. E. King, H. S. Mason, and M. Morrison, eds.), Vol. 2, John Wiley and Sons, Inc., New York, 1964: on conformational change and coupling.

Chance, B. C., and Mela, L., *Nature* **212**, 372 (1966): an evaluation of the experimental foundations of the chemiosmotic hypothesis.

Green, D. E., and Hatefi, Y., *Science* **133**, 3445 (1961): on the complexes of the electron transfer chain.

Green, D. E., Asai, J., Harris, R. A., and Penniston, J. T., *Arch. Biochem. Biophys.* **125**, 684 (1968a): on the conformational basis of energy transformations in mitochondria.

Green, D. E., Haard, N. F., Lenaz, G., and Silman, H. I., *Proc. Natl. Acad. Sci. U.S.* **60**, 277 (1968b): on the noncatalytic proteins of membrane systems.

Harris, R. A., Penniston, J. T., Asai, J., and Green, D. E., *Proc. Natl. Acad. Sci. U.S.* **59**, 830 (1968): on the correlation between configurational change and the energized state.

Hatefi, Y., *Advan. Enzymol.* **25**, 275 (1963): on the complexes of the electron transfer chain which react with coenzyme Q.

Korman, E. F., VandeZande, H., and Green, D. E., In preparation: on the resolution of Complex IV into catalytic and noncatalytic proteins.

Margolis, S. A., Baum, H., and Lenaz, G., *Biochem. Biophys. Res. Commun.* **25**, 133 (1966): inhibition of energy transformation in deuterium oxide.

Mitchell, P., *Nature* **191**, 144 (1961); *Biol. Rev. Cambridge Phil. Soc.* **41**, 445 (1966): on the chemiosmotic theory of coupling.

Moore, C. L., *Biochemistry* **7**, 300 (1968): an evaluation of the chemiosmotic hypothesis.

Penniston, J. T., Harris, R. A., Asai, J., and Green, D. E., *Proc. Natl. Acad. Sci. U.S.* **59**, 624 (1968): on the conformational basis of energy transformations in mitochondria and membrane systems generally.

Silman, H. I., Rieske, J. S., Lipton, S. H., and Baum, H., *J. Biol. Chem.* **242**, 4687 (1967): on the noncatalytic protein component of Complex III of the mitochondrial electron transfer chain.

Urry, D. W., and van Gelder, B. F., *Symp. Cytochromes, Osaka 1967*, p. 58; University of Tokyo Press, 1967: changes in circular dichroism during electron transfer in Complex IV.

Utsumi, K., and Packer, L., *Arch. Biochem. Biophys.* **121**, 633 (1967): conformational changes in oxidative phosphorylation.

KEY REFERENCES FOR CHAPTER VI

Complexes of the electron transfer chain as the units of coupling
Green and Hatefi (1961), Hatefi (1963)

Conformational changes in Complex III
Baum *et al.* (1967), Green *et al.* (1968b), Silman *et al.* (1967) Urry and van Gelder (1967)

Discontinuous electron transfer
Baum *et al.* (1967), Urry and van Gelder (1967)

Conformational basis of energy transduction
Boyer (1964), Green *et al.* (1968a), Harris *et al.* (1968), Penniston *et al.* (1968), Utsumi and Packer (1967)

Chemiosmotic hypothesis
Chance and Mela (1966), Mitchell (1966), Moore (1968)

Pseudoenergized swelling
Blondin *et al.* (1969)

Multiple modes of energy utilization
Green *et al.* (1968a)

Transmission of the energized state
Green *et al.* (1967a), Harris *et al.* (1968)

CHAPTER VII

UTILIZATION
AND MANIPULATION
OF CONFORMATIONALLY
CONSERVED ENERGY

The energy transformations of mitochondria can be summarized satisfactorily by the simple diagram shown below:

$$\text{Electron transfer} \underset{}{\overset{I}{\rightleftharpoons}} \text{[energized state]} \underset{}{\overset{II}{\rightleftharpoons}} \text{ATP}$$
$$\downarrow$$
$$\text{work performances}$$

There are two functionally different systems capable of generating the energized state of the repeating units in the inner membrane: I, one or another complex of the electron transfer chain; and II, the ATPase complex. System I is localized in the basepieces; System II in the headpiece-stalk sectors. The energized state is the same whether generated by System I or System II. The electron transfer process in the basepieces, or the coupled hydrolysis of ATP in the headpiece, initiates the generation of the energized state. Whatever the nature of the resulting conformational change in the repeating unit, and wherever in the repeating unit this change is initiated, the ultimate nature of the energized state is independent of the mode of generation.* Topologically, Systems I and II are in a right angle relation to one another. The electron transfer process proceeds in the plane of the membrane whereas the coupled hydrolysis of ATP takes place in a direction orthogonal to the membrane. The stalk, which is also

* As we shall see below, however, the precise configurational form of the energized state, although independent of the initial source of energy, is dependent upon local environmental factors.

orthogonal to the basepiece, must be presumed to be the functional connection between changes in the headpiece and changes in the basepiece.

As indicated by the diagram, the generation of the energized state whether by electron transfer or by hydrolysis of ATP is a reversible process. Thus, the energized state can drive electron transfer in the direction opposite to the normal flow of electrons, and it can be utilized for the synthesis of ATP by the union of ADP and P_i. All other modes of utilization of the energized state appear to be irreversible, at least under the conditions that are commonly employed for their demonstration. These various modes of utilization are encompassed within the category of work performances.

At least three distinct work performances that can be powered by the energized state have been described. These are, respectively: (1) energized transfer of a hydride ion from DPNH to TPN^+; (2) energized translocation of divalent ions and inorganic phosphate; and (3) energized translocation of monovalent ions with accompanying swelling. We may consider each of these work performances as alternative devices for discharging the energized state.

CONFORMATIONAL CHANGES IN REPEATING UNITS LEADING TO
CONFIGURATIONAL CHANGES IN THE INNER MEMBRANE

There are three configurations which the inner membrane can assume: the nonenergized, the energized, and the energized-twisted configurations. The cristae of the mitochondrion* are sacs having the general shape of a hot-water bottle. The sacs are aligned in parallel arrays orthogonal to the limiting outer membranes (see Figure IV.1), and their open ends are fused into the inner boundary membrane of the outer membrane system. In the nonenergized state the sacs are flattened so that the opposing membranous walls of each sac are parallel. The lumen within the sac is continuous with the lumen between the two boundary membranes of the outer membrane system. When the repeating units of the inner membrane are energized as a result of electron transfer, the flattened sac of the crista bulges

---

* Not all mitochondria have the same ultrastructure of the cristael membranes. The description given in the text is characteristic of beef heart mitochondria and applies to most mitochondria except for adrenal and sperm mitochondria.

Fig. VII.1. The configurations of the crista in heart muscle *in situ*: nonenergized (A), energized (B), and zigzag (energized-twisted) (C). Taken from Penniston *et al.*, *Proc. Natl. Acad. Sci. U.S.* **59**, 624 (1968).

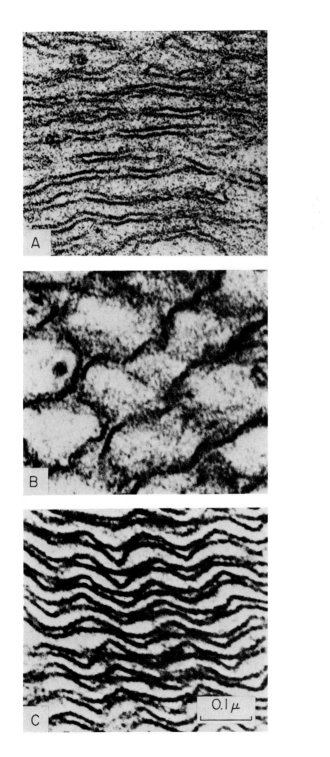

A

B

C

0.1 μ

out (like a full hot-water bottle) and in cross section appears vesicular (see Figure VII.1). There is yet another configuration which the membrane can assume when the repeating units are energized; this is achieved in presence of inorganic phosphate. This condition leads to a zigzag arrangement of the cristae in cross section (see Figure VII.1). Transitions from any one of these three configurational states of the membrane to another state are reversible, and under favorable conditions, are almost as fast as the turnover of the electron transfer components.

The configuration of the membrane is an expression of the way in which the repeating units nest together. When the basepieces of the repeating units are cylindrical or cuboidal, the nesting faces are parallel and the membrane has, thus, little or no curvature (see Figure VII.1). The basepieces of the repeating units in the nonenergized state have a symmetrical geometry. The parallel arrangement of the opposing walls of each crista is an expression of the nesting of symmetrical repeating units. When the repeating units are energized by oxidation of substrate or by hydrolysis of ATP, the basepieces become more asymmetric in shape (nesting faces are not parallel) and the cristae then assume a more nearly spherical configuration in accommodation to this change in shape of the repeating units (see Figure VII.1). Finally, when the repeating units are energized in the presence of inorganic phosphate, the basepieces assume a more twisted geometry because of attraction between headpieces, and this twist in the repeating units is translated into a zigzag configuration of the cristae.

A single crista is never detected in a mixed configuration. Presumably, there must be transitional configurations, but the configurational changes are so rapid that the probability of fixing a crista in an "in between" configuration is extremely small.

The three configurational states of the inner membrane we have described above are readily observed in mitochondria when *in situ*. The parallel and regular alignment of cristae which obtains *in situ* makes possible the visualization of orderly domains of cristae— domains in which cristae are found in the same configuration. When mitochondria are isolated by standard procedures, the regularity of the arrays of cristae is lost (the *in situ* arrangement of the crista can, however, be achieved under appropriate experimental conditions). At present there is insufficient evidence to justify an attempt at a full listing of the forces which are responsible for the disarray of cristae in isolated mitochondria. Sufficient to say, there is a tendency of cristae to aggregate (see Figure VII.2), and this apposition of one crista with its neighbor complicates enormously the electron microscopic visualization and interpretation of the configurational changes which occur in the cristae of isolated mitochondria. Further complication is introduced by the tendency of apposed cristae to comminute (see Figure VII.2). If, however, we bear in mind the

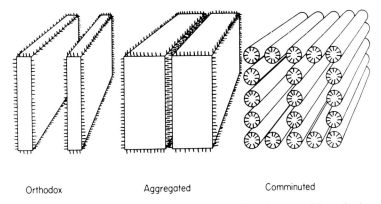

Orthodox   Aggregated   Comminuted

Fig. VII.2. Diagrammatic representation showing the apposition of cristae leading to engagement of repeating units and to tubularization. Taken from Penniston *et al., Proc. Natl. Acad. Sci. U.S.* **59**, 624 (1968).

fact that the same basic changes in repeating units take place whether the mitochondrion is *in situ* or separated from the cell, then the picture becomes much clearer. What is different is not the essential changes in geometry of the repeating units in transition from the nonenergized to the energized state and finally to the energized-twisted state; rather it is the way in which the membrane accommodates to these conformational changes in the repeating units which is different.

    The arrangement of cristae in mitochondria *in situ* will be referred to as the *orthodox* mode; the three configurations of the membrane of the crista (nonenergized, energized, and energized-twisted) have already been described above. In isolated mitochondria, the cristae are usually in the *aggregated* mode, i.e. the walls of neighboring cristae expand to the point of contact. In this mode, all of the three

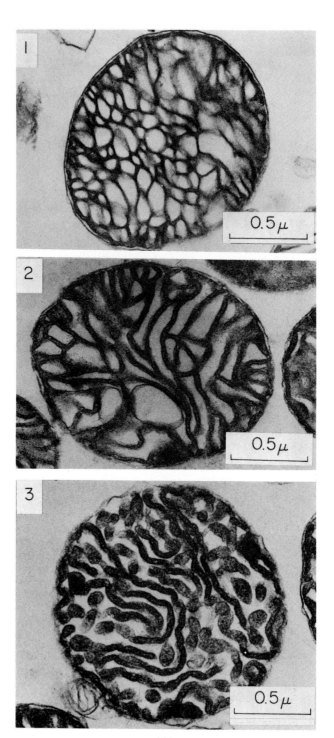

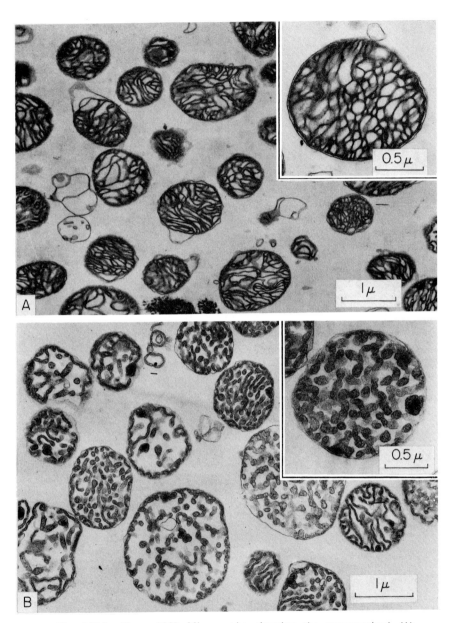

Fig. VII.3. [Page 100] Micrographs showing the nonenergized (1), energized (2), and energized-twisted (3) configurations of the crista of beef heart mitochondria in the aggregated mode. [Page 101] Electron micrographs showing a field of mitochondria in the nonenergized configuration (A) and in the energized-twisted configuration (B).

configurations of the membrane are shown in Figure VII.3. Although it is a complex matter to relate in geometric terms the nonenergized and energized configurations in the aggregated mode to the corresponding configurations in mitochondria *in situ*, one may say confidently that the repeating units, regardless of the configurational mode, are collapsed in the nonenergized conformation and extended in the energized state. The final expression of these conformations of the repeating units can be profoundly modulated by osmotic and other forces.

The energized-twisted configuration of the membrane of the cristae in isolated mitochondria arises by a process of tubularization. Almost invariably the aggregated cristae in the energized-twisted state are fragmented into snakelike tubules (see Figure IV.7). This configuration is referred to as the *comminuted* mode of the crista. In general we shall be concerned principally with the following configurational transitions: from nonenergized in the aggregated mode to energized and from energized to energized-twisted in the comminuted mode.

When isolated mitochondria are exposed to Zn^{2+} or Ca^{2+} or to the endotoxin of *Bordetella*, the cristae rearrange and assume the regular arrays characteristic of cristae in mitochondria when *in situ*. Moreover, the same three configurational states observed *in situ* are now found in the cristae of such treated mitochondria when energizing conditions are imposed.

GENERATION OF THE ENERGIZED STATE

The energized state can be generated by the electron transfer process or by hydrolysis of ATP. Inhibitors of the electron transfer process such as antimycin will suppress the generation of the energized state by electron transfer, but not the generation of the energized state by ATP hydrolysis. Conversely, inhibitors of ATP hydrolysis such as rutamycin will suppress the generation of the energized state by hydrolysis of ATP, but not the generation by electron transfer. Atractyloside, which prevents phosphoryl transfer from external ATP to internal ADP, has the same effect as rutamycin in suppressing the ATP-induced energized state.

Any substrate of the electron transfer process can induce the energized state (DPNH, succinate, reduced cytochrome c, QH_2). When reduced cytochrome c is the electron donor, antimycin A* has no effect on the generation of the energized state but cyanide or azide* is effective. Similarly rotenone* inhibits the DPNH-generated energized state but does not inhibit the energized state generated by succinate.

The energized state includes both the energized and energized-twisted configurations of the membrane of the crista. In absence of inorganic phosphate, the energized state is formed; in presence of orthophosphate (0.5 mM or higher), the energized-twisted state is formed. Inhibitors of electron transfer or of ATP hydrolysis prevent the formation of both energized states under the conditions specified above.

GENERATION OF THE ENERGIZED-TWISTED STATE

Not only phosphate but also arsenate can induce the transition of the energized to the energized-twisted state, but the level of inorganic arsenate required is about ten times higher than that of inorganic phosphate. Monovalent anions, such as acetate and propionate, can induce the transition from energized to energized-twisted state but again the concentrations required are higher than that of inorganic phosphate. Appropriate salts can induce a pseudoenergized-twisted state in sucrose-free media in absence either of electron transfer or of ATP. Inhibitors of electron transfer, or of ATP hydrolysis, have no influence on the salt-induced generation of the pseudoenergized-twisted state. The pseudoenergized-twisted state leads to swelling, and in this respect the energized-twisted and psuedoenergized-twisted states are equivalent.

DISCHARGE OF THE ENERGIZED STATE

There are four known ways of discharging the energized state: (1) by uncouplers; (2) by ADP (in presence of phosphate); (3) by divalent metal ions such as Ca^{2+} (in presence of phosphate); and (4) by monovalent salts such as ammonium acetate. Discharge via

* Rotenone inhibits electron transfer through Complex I; antimycin A, through Complex III; and cyanide or azide, through Complex IV.

routes 2, 3, and 4 leads, respectively, to ATP synthesis, to translocation of Ca^{2+} and phosphate, and to swelling. Discharge via uncouplers prevents any work performance from taking place.

It can hardly be a coincidence that the very reagents which can discharge the energized state are precisely the ones which determine the various uses to which the energized state can be put. Thus, ADP can serve as phosphoryl acceptor in oxidative phosphorylation; Ca^{2+} can serve as a partner for the phosphate ion in the deposition of $Ca_3(PO_4)_2$ during translocation; and monovalent salts can serve as the determinants of energized swelling. From this relationship it would follow that there is a cycle first of generation of the energized state and then of discharge. Thus, in the generative phase of oxidative phosphorylation, the energized-twisted state induced in presence of orthophosphate leads to the formation of a conformationally constrained phosphate group with the chemical potential of a phosphoryl group; in the discharge phase of the cycle this phosphate group is transferred to ADP with synthesis of ATP. In translocation of divalent metal ions, the phosphate group is transferred to Ca^{2+}, or to some other suitable divalent metal ion (Me^{2+}), with deposition of $Me_3(PO_4)_2$. In energized swelling, the anion (probably in the form of the undissociated acid) induces the energized-twisted state; in turn this state is discharged by the cation* with translocation of both anion and cation into the interior space of the inner membrane tubule. This translocation is preliminary to a conformational rearrangement of the inner membrane which leads to swelling.

SPEED OF CONFIGURATIONAL CHANGE

If the conformational changes expressed as configurational alteration of the membranes of the cristae are, indeed, the primary events in energized coupling, it would follow that these changes should be as rapid as the turnover of the components of the electron transfer chain. In so far as rates can be determined either by electron microscopic examination of rapidly fixed samples or by the light scattering

---

* In some cases the monovalent cation can gain access to the basepiece only in the presence of a facilitating agent, such as valinomycin; the ammonium ion is not restricted in this way. The facilitation phenomenon is discussed later in this chapter.

changes which accompany conformational change, it would appear
that the configurational changes are sufficiently rapid to be of the
same order of magnitude as the rates of turnover of the catalytic
components of the electron transfer chain. There appears, therefore,
to be no kinetic reason for considering these conformational changes
as other than the primary events.

MECHANISM OF ENERGIZED TRANSHYDROGENATION

Energized transhydrogenation involves a cycle first of generating
the energized state by either substrate or ATP and then of discharging
the energized state by TPN$^+$ (in the presence of DPNH). One of the
complexes of the inner membrane catalyzes the transfer of a hydride
ion from TPNH to DPN$^+$. This reaction reaches equilibrium when the
value of the ratio

$$\frac{[\text{DPNH}] \times [\text{TPN}^+]}{[\text{DPN}^+] \times [\text{TPNH}]}$$

is about 1. When the transfer of a hydride ion from DPNH to TPN$^+$
is energized, the equilibrium value for the hydride transfer, i.e.

$$\frac{[\text{DPN}^+] \times [\text{TPNH}]}{[\text{DPNH}] \times [\text{TPN}^+]}$$

is close to 500 and the reaction involving transfer of a hydride ion
from TPNH to DPN$^+$ is excluded. It would appear that when the
transhydrogenating complex is energized in its reduced form (generated by DPNH), one of the functional groups of this complex becomes
extremely negative in reducing potential, and this conformationally
induced species with high reducing power and high specificity for
TPN$^+$ as oxidant, can reduce TPN$^+$ almost completely. In this
instance the conformational change inherent in the energized process
leads to a change in the oxidation-reduction potential of a functional
group in the energized complex (probably an internal hydride
acceptor) and the interaction of this reducing group with the external
hydride acceptor, TPN$^+$, leads to discharge of the energized state.
There is, thus, a 1:1 relation between the number of molecules of
DPNH that have been oxidized during energized transhydrogenation
and the number of molecules of TPN$^+$ reduced. When the transhydrogenation complex becomes energized, then the specificity for

the direction of transhydrogenation changes dramatically. The hydride transfer from TPNH to DPN^+ is excluded, and only the transfer reaction from DPNH to TPN^+ is favored.

FACILITATED ENTRY OF METABOLITES

Energized permeation of mitochondria by substrate molecules, such as malate, oxaloacetate, succinate, is a phenomenon that has not as yet been thoroughly studied. Conformational change of the inner membrane is apparently the determinant of the change in the permeation of the mitochondrial outer membrane by dicarboxylic acids. The rearrangement of the mitochondrial inner membrane during the energized process undoubtedly has a profound influence on the properties of the outer membrane* and, in particular, affects the permeation of the repeating units of the outer membrane by dicarboxylic acids.

CAN MORE THAN ONE WORK PERFORMANCE BE CARRIED ON SIMULTANEOUSLY?

Under conditions that maximize translocation of Ca^{2+} and inorganic phosphate, oxidative phosphorylation does not proceed at all. Apparently Ca^{2+} is far more efficient that is ADP in discharging the energized-twisted state induced by inorganic phosphate. However, translocation of Mg^{2+} plus phosphate ion is considerably slower than that of Ca^{2+} plus phosphate ion. ADP can compete favorably with Mg^{2+} in discharging the energized-twisted state; thus, both oxidative phosphorylation and translocation of Mg^{2+} can proceed simultaneously. This is a permissible option because the same energized state (phosphate-induced, energized-twisted state) is common to both processes.

The energized-twisted state induced by the combined presence of salt (the anion of which is not inorganic phosphate), substrate, and an appropriate facilitating agent, such as valinomycin, cannot be discharged by ADP. We must assume, therefore, that the energized-

---

* This must, in particular, be the case at the points of fusion between the cristae and the inner boundary membrane of the outer membrane.

twisted state induced under the above set of conditions is distinct from the analogous state induced by the combined presence of inorganic phosphate and substrate. No competition can develop between energized swelling and oxidative phosphorylation because these two processes involve different forms of the energized state. Once the energized state in the form induced by anions other than phosphate is established, then the option for oxidative phosphorylation is no longer open. Similarly, once the energized state in the form induced by inorganic phosphate is established, then discharge by ADP leading to oxidative phosphorylation takes precedence over discharge by monovalent cations leading to energized swelling.

Swelling may also be induced by appropriate salts when mitochondria are suspended in sucrose-free media. This induction is independent of electron transfer or ATP hydrolysis. The configurational state of the inner membrane in mitochondria thus treated, corresponds closely with that of the energized-twisted state. But, despite the formal resemblance in gross ultrastructural appearance, there is one property which is profoundly different. Uncouplers discharge the phosphate-induced energized-twisted state or the analogous energized-twisted state induced by salt in presence of substrate plus either valinomycin or gramicidin; but, uncouplers do not discharge the configurationally equivalent state induced by salt in sucrose-free media. Thus, we have three variants of the energized-twisted state; one dischargeable by ADP, by divalent cations, or by uncouplers; the second dischargeable by monovalent ions or by uncouplers; and a third variant that is not dischargeable by ADP, salts, or uncouplers. It is not excluded, however, that what is different is not the configurational state but the conditions of formation. Once these conditions are changed, the energized-twisted state may be dischargeable in all cases by uncoupler.

THE ATPASE COMPLEX

The complex which reversibly couples the hydrolysis of ATP to the generation of the energized state may be defined as the ATPase complex in its most complete form. This complex would include the set of associated proteins in the headpiece, which collectively hydrolyzes ATP, and also the stalk, which is concerned both with the coupling of this hydrolysis to the conformational change, and also with the transmission of the conformational change from the headpiece to the basepiece. The ATPase activity of the headpiece is rutamycin-insensitive whereas the ATPase activity of the headpiece-stalk sector is rutamycin-sensitive. Neither the isolated headpiece nor the isolated headpiece-stalk sector shows evidence of conformational change associated with hydrolysis of ATP. This coupling of ATP hydrolysis to conformational change (or at least to a change which does not spontaneously discharge) appears to require the link of the headpiece-stalk sector to the basepiece.

When the headpiece-stalk sector of the inner membrane repeating unit is isolated in a form free of phospholipid, no ATPase activity is measurable. When, however, the headpiece-stalk unit is allowed to interact with micellar phospholipid, the ATPase activity is fully reconstituted. When not attached to the stalk, the headpiece is a soluble species, the ATPase activity of which is rutamycin-insensitive and unaffected by phospholipid.

The ATPase complex (headpiece-stalk sector of the repeating unit) interdigitates with the electron transfer system in such a way that the conformational change induced by ATP hydrolysis can be translated into the same conformational state that is induced by electron transfer in the basepiece. Conversely, the energized state induced by electron transfer can be channeled into a reversal of the conformational change by which ATP is hydrolyzed.

UNCOUPLERS

A sizeable list of compounds is known which not only can uncouple electron flow or ATP hydrolysis from the generation of the energized state, but also can "deenergize" the already formed energized state. The most powerful reagent in this list is a derivative of carbonyl-cyanide phenylhydrazone. Two other well known uncouplers are 2,4-dinitrophenol and dicumarol.

A phenomenon which is central to the action of uncouplers is the transfer of the energized state from basepiece to basepiece by a type of exchange reaction. This phenomenon underlies the fact that the entire membrane of a crista can be energized by electrons originating in only one of the multiple species of basepieces (electron flow through the other species being excluded). This phenomenon also underlies the fact that highly efficient uncouplers such as carbonyl-cyanide phenylhydrazone can completely uncouple oxidative phosphorylation even at concentrations which are far below those required for a 1:1 molecular ratio of uncoupler to repeating unit (less than one molecule of uncoupler per 50 or more repeating units). If the uncoupler combines stoichiometrically with one repeating unit at the appropriate site and if the energized state is rapidly dis-charged in presence of the uncoupler, it is possible for the repeating unit in question to serve as a sink for the energized state.* The transfer of the energized state will eventually lead to the discharge of the energized state in the entire membrane via the repeating unit which is combined with the uncoupler. It would follow that repeating units in a membrane which are uncoupled by virtue of structural damage would act in the same way as uncouplers, i.e. these would serve as sinks for the energized state. Dislocations of the repeating units induced by lyophilization and also by freezing and thawing have, indeed, been shown to lead to the uncoupling of the cristae of the inner membrane.

There is an all-or-none component in the energizing of the inner mitochondrial membrane. The membrane is either energized or not energized; it is never found in an intermediate state. It is as if the membrane of a crista were energized or discharged in waves to avoid mixed states and, consequently, mixed geometries, which could lead to considerable strain on the membrane.

RESPIRATORY CONTROL AND THE POISING OF THE ENERGIZED STATE

Mitochondria have a measure of stability in the energized state. Thus, in presence of substrate and inorganic phosphate, well coupled mitochondria respire at a reduced rate until external ADP is added. Respiration reaches the maximal rate as soon as the energized state can be discharged by external ATP formation (or by uncouplers or

---

* The efficiency of such a sink would clearly depend upon the rate of generation of the energized state, so that at high rates of respiration, a steady-state level of energized basepieces might be established which would allow some phosphoryla-tion of ADP to proceed. This is what is actually observed in presence of low concentrations of uncoupler.

by work performances). The phenomenon of respiratory control is an expression of the poising of the mitochondrion in the energized or energized-twisted state. If the stability of this state were infinite, no respiration would take place. But, the stability at 30° is only partial. The respiratory control ratio (the rate of respiration in presence of ADP divided by the rate in absence of ADP) is at most 10–12, and usually considerably less. When mitochondria are comminuted to submitochondrial particles, respiratory control declines dramatically. Apparently, the energized or energized-twisted state is far more unstable in submitochondrial particles than in untreated mitochondria. When the submitochondrial particles are vesicular, respiratory control is virtually zero. However, a significant amount of respiratory control is found in submitochondrial particles which retain the tubular character of the crista. This correlation points to the importance of the geometry of the membrane in the stabilization of the energized state. The fact that particles with little or no respiratory control can synthesize ATP quite efficiently under appropriate conditions implies that the spontaneous discharge process is less rapid than the second transduction (headpiece-stalk mediated synthesis of ATP). The site of thermal discharge in such particles may actually be the headpiece-stalk connection and not the basepieces themselves, since rutamycin (which inhibits the transfer of the energized state between basepiece and headpiece-stalk) can reimpose a degree of respiratory control upon phosphorylating submitochondrial particles.

THE DELIVERY OF PHOSPHORYL GROUPS FROM INTERNALLY GENERATED ATP TO EXTERNAL ADP

In the intact mitochondrion there is a major logistic problem involved in the delivery of ATP generated in the inner membrane to systems external to the mitochondrion. The first principle that had to be learned was that the inner membrane is essentially impenetrable to adenine nucleotides. Thus, it is impossible to deliver ATP as such, i.e. in the form of the whole molecule. From a study of the effect of the plant glycosidic steroid, atractyloside, on oxidative phosphorylation in intact mitochondria, it became obvious that the delivery problem was solved by transferring phosphoryl groups, not by trans-

ferring adenine nucleotides. The set of enzymes which participates
in this delivery is our immediate concern.

The gross pathway by which a phosphoryl group of ATP generated
in the headpiece of the repeating units of the inner membrane is transferred externally may be described as follows: from the cristael membrane to the structured complex in the lumen (intermembrane system),
from intermembrane system to outer boundary membrane, and from
outer boundary membrane to external ADP. There may be no transfer
of phosphoryl groups as such within the cristael membrane. The equilibrium between the energized state and ATP would make unnecessary
intramembrane transfer of phosphoryl groups. Three phosphoryl
transfers would have to take place—from ATP in the cristael membrane to a phosphoryl acceptor in the intermembrane system, from
intermembrane system to a phosphoryl acceptor in the outer boundary
membrane, and from boundary membrane to external ADP. The
transfer of a phosphoryl group may be from ATP to ADP, from ATP
to GMP, from GTP to AMP, etc. The enzyme is the determinant of
both the phosphoryl donor and the phosphoryl acceptor. The atracyl-
ate-sensitive enzymes which catalyze phosphoryl transfers from cristael membrane to intermembrane system, and from intermembrane
system to outer boundary membrane, presumably must be vectorially
aligned with respect to these three systems to achieve these transfers.
Thus, the phosphoryl transfer takes place within the molecular tracks
provided by the transferring enzyme. There is no generalized movement of phosphoryl groups; rather, there is enzyme-directed movement. This problem of phosphoryl transfer, although less complex for
submitochondrial particles (which lack an outer membrane and have
inner membrane headpieces directed outward), is nonetheless applicable because there is need to transfer a phosphoryl group from the
repeating unit to ADP that is external to the particle.

FACILITATION OF SALT-INDUCED ENERGIZED SWELLING BY GRAMICIDIN
AND VALINOMYCIN

Salts of monovalent ions at a concentration of 0.01 M can induce
energized swelling of mitochondria against an osmotic gradient
imposed by a nonpenetrating solute such as sucrose. This swelling
requires energizing by either substrate or ATP; it requires also that
the anion of the monovalent salt be a weakly dissociated acid (acetate,

propionate, phosphate, formate, etc.); and, finally, it requires the presence of a facilitating agent such as gramicidin or valinomycin. It is the role of the facilitating agents for energized swelling that is our present concern. These agents are polypeptides of low molecular weight, and their action is highly specific. The rationale of this facilitation appears to be the following. The entry of both cation and anion into the interior of the membrane to appropriate sites therein is a necessary condition for swelling to take place. Gramicidin and valinomycin are reagents which can dissolve in the membrane-bound phospholipid phase or in the hydrophobic sectors of the membrane repeating units. The geometry of these molecules is such that they can contain ion pairs such as K^+ and chloride. The analogy of a pouch at the molecular level is helpful in visualizing the mechanism of facilitation by gramicidin and valinomycin. In a sense the antibiotics in question act as devices for transport of ions into the nonpolar regions of a membrane. The facilitation of entry by gramicidin is general for all alkali metal salts and is not specific for particular cations, although some cations are more effective than others. The facilitation of entry by valinomycin is more effective for potassium salts than for salts of other monovalent metals. Salts, such as ammonium acetate, which are soluble in nonpolar media readily penetrate the mitochondrion, and this penetration does not require facilitation.

Divalent ions such as Cd^{2+}, Zn^{2+} and Ca^{2+} can duplicate the facilitation of entry of monovalent salts achieved by gramicidin and valinomycin. However, this facilitation in the case of Zn^{2+} and Ca^{2+} requires an initial energy-dependent translocation of the divalent metal. No such requirement applies to facilitation by gramicidin, valinomycin, and Cd^{2+}.

H. A. Lardy has described a series of antibiotics the members of which reverse the flow of monovalent ions accompanying energized swelling. In the presence of these antibiotics, the cations concentrated during the swelling cycle leave the mitochondrion. These antibiotics duplicate the effect of uncouplers which discharge the energized state. One of the most potent members of the Lardy series (Nigericin) has been found to be a powerful uncoupler of photophosphorylation. Nigericin also has the capability for binding anions such as chloride and for facilitating transfer of chlorides across the mitochondrial membrane.

CONFIGURATIONAL CHANGES DURING ENERGIZED SWELLING

During swelling of mitochondria induced by 0.01 M salt under energizing conditions, or by salt under nonenergizing conditions (i.e. absence of substrate, or ATP, and in absence of any nonpermeant solute), a unique pattern of ultrastructural changes is recognizable by electron microscopy (cf. Figure VII.4). First, the energized-twisted state (or its pseudoenergized equivalent) is established. Then the tubules swell and fill out to form vesicles. The small vesicles coalesce to form larger vesicles; ultimately one large spherical membrane is formed. The volume of water contained by this single spherical membrane is some five times greater than the volume of water in the cristae of the same mitochondrion prior to swelling.

Berton Pressman has shown that mitochondria which have undergone energized swelling in presence of salts can shrink back spontaneously to the unswollen state. This cycle of change can be repeated four or more times. What accounts for the oscillatory behavior of the mitochondrion? It would appear that when mitochondria swell, coupling of electron transfer to the induction of the energized state becomes progressively more defective. When a critical point in this progressive uncoupling is reached, the balance is upset and the process of swelling reverses. Water and ions leave the mitochondrion. As the repair process continues, the coupling capability is reacquired and a new cycle of swelling is initiated. The oscillatory behaviour of mitochondria in respect to the swelling-shrinking cycle is really an oscillation in respect to coupling capability which wanes in the swollen mitochondrion and recovers in the nonswollen mitochondrion.

ENERGIZED TRANSLOCATION OF DIVALENT METAL IONS

Energized swelling is invariably a consequence of the substrate- (or ATP-) induced accumulation of salts within the lumen of the tubules of the inner mitochondrial membranes. The competent salts are monovalent species. However, when mitochondria are exposed to divalent metal salts as well as substrate and inorganic phosphate, no swelling is observed and the containment of divalent cation takes an entirely different form. The divalent cation (Me) is deposited with phosphate as an insoluble salt with the composition of $Me_3(PO_4)_2$. In energized translocation of divalent cations, the

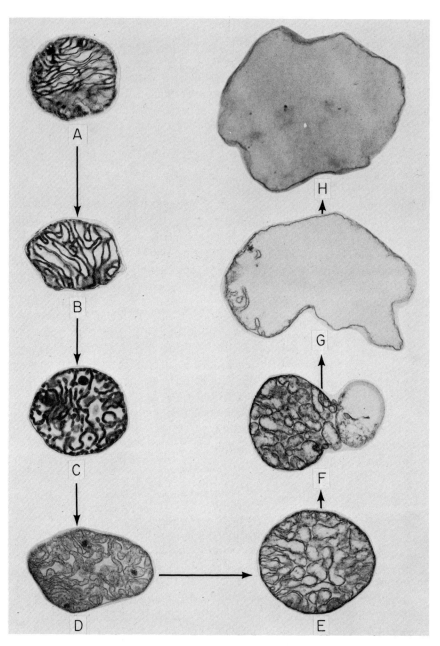

Fig. VII.4. Diagrammatic representation showing the ultrastructural changes during swelling of mitochondria. (A) initial state—nonenergized configuration in the aggregated mode; (B) energized configuration in the aggregated mode; (C) energized-twisted configuration in the aggregated mode; (D) swollen forms of the energized-twisted cristae; (E) vesicular forms of the energized-

ratio of cation to anion translocated and deposited is thus constant
(3:2). But, this phenomenon is restricted to phosphate or arsenate
as anions. There is no limitation in amount when Ca^{2+} plus P_i are
translocated by mitochondria in presence of substrate, since the
amount of insoluble salts which may be deposited has no practical
limit.

The stoichiometry of the energized accumulation of calcium in
presence of phosphate has been intensively studied by several in-
vestigators. It would appear that discharge of the energized state
of a single repeating unit leads to the deposition of two calcium ions.
Each such discharge permits the cycle of energizing and deenergizing
to be repeated once again.

ENERGIZED MOVEMENT OF PROTONS

The formulation of Mitchell's theory of oxidative phosphorylation
has inspired an intensive study of proton movements associated with
energy transformation in mitochondria. Unfortunately, it is not yet
possible fully to interpret the experimental observations. This is
hardly surprising since there is scarcely a process in biochemistry
which, if fully described in molecular terms, does not involve net
proton movements. Since many processes in the mitochondrion are
vectorial and take place in association with anisotropic membranes,
it would be expected that such processes would be characterized by
vectorial proton movements. A proportion of the energy conserved in
a conformationally perturbed complex might indeed be related to
changes in ionization of specific groups in the proteins of the complex.
To that extent Mitchell's "proton motive force" (see Chapter VI)
might be a way of describing the conformational energy of a membrane
in the energized state.

SUBSTRATE LEVEL PHOSPHORYLATION

The α-ketoglutarate dehydrogenase can synthesize ATP by the
coupling of pyrophosphate bond formation to the enzymic oxidation
of α-ketoglutarate. It may be instructive to compare ATP synthesis
via substrate level phosphorylation with ATP synthesis via oxidative

---

twisted cristae; (F) coalescence of the vesicular membranes; (G) terminal
state in the coalescence of vesicles to form one expanded vesicular membrane
apart from a few residual orthodox cristae; (H) swelling at its most complete
form.

phosphorylation. But, first let us review the essential events in substrate level phosphorylation:

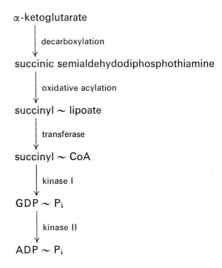

α-ketoglutarate

decarboxylation

succinic semialdehydodiphosphothiamine

oxidative acylation

succinyl ∼ lipoate

transferase

succinyl ∼ CoA

kinase I

GDP ∼ P_i

kinase II

ADP ∼ P_i

The oxidation of α-ketoglutarate (more precisely the oxidation of succinic semialdehydodiphosphothiamine) by oxidized lipoate leads to the formation of a succinyl derivative of reduced lipoate. The link between the succinyl group and one of the two sulfur atoms of reduced lipoate (thioester link) is a "high energy" link. The free energy released in the oxidation is conserved by the formation of succinyl lipoate. Two kinases then intervene for the transfer of the bond thus established between succinyl and lipoate residues on the one hand to the pair of residues ADP and P_i, on the other hand. Succinyl-CoA and GDP-P_i are the intermediates in this transfer sequence. The kinase which transfers the bond reversibly from succinyl-CoA to GDP-P_i interacts with GTP to form phosphoryl enzyme and GDP. The amino acid residue in the kinase which is phosphorylated has been identified by Boyer as a histidine residue.

In oxidative phosphorylation associated with the electron transfer chain the energy of oxidation is conserved as conformational energy and then transformed into the bond energy of ATP. Substrate level phosphorylation is another kind of oxidative phosphorylation in which the transformation is direct. The oxidation is directly linked to the formation of a high energy bond. In line with this essential difference the outer membrane plays no role in energizing the work performances of the mitochondrion—a role which requires the generation of the energized state.

SELECTED REFERENCES

BOOKS

Chance, B., ed., "Energy-Linked Functions of Mitochondria," Academic Press, New York, 1963: a useful compendium on energized ion movements in mitochondria.

SPECIAL ARTICLES

Allmann, D. W., Harris, R. A., and Green, D. E., *Arch. Biochem. Biophys.* **122**, 766 (1967): on phosphoryl transfer reactions in mitochondria.

Asai, J., Asbell, M. A., Harris, R. A., and Green, D. E., in preparation: form and function of repeating units.

Asai, J., Blondin, G., Vail, J., and Green, D. E., *Arch. Biochem. Biophys.* in press (1969).

Beyer, R. E., *Arch. Biochem. Biophys.* **123**, 41 (1968): reconstitution of oxidative phosphorylation in submitochondrial particles by a soluble phosphoryl transferase.

Blondin, G., Vail, W. J., and Green, D. E., *Arch. Biochem. Biophys.* **129**, 158 (1968): potentiation of pseudoenergized swelling by Ca^{2+}, Zn^{2+}, and Cd^{2+}.

Brierley, G. P., Bachmann, E., and Green, D. E., *Proc. Natl. Acad. Sci. U.S.* **48**, 1928 (1962): on the active transport of inorganic phosphate and magnesium ions by beef heart mitochondria.

Brierley, G. P., Bhattacharyya, R. H., and Walker, J. G., *Biochem. Biophys. Res. Commun.* **24**, 269 (1966): induction of K^+ transport in isolated heart mitochondria by zinc ions.

Green, D. E., Asai, J., Harris, R. A., and Penniston, J. T., *Arch. Biochem. Biophys.* **125**, 684 (1968): on the conformational basis of energy transductions in mitochondria.

Griffiths, D. E., and Haslam, J. M., *Biochem. J.* **104**, 52P (1967): on the factors affecting the penetration of oxaloacetate and L-malate into rat liver mitochondria.

Hackenbrock, C. R., *J. Cell Biol.* **30**, 269 (1966): ultrastructural basis for metabolically linked mechanical activity in mitochondria.

Harris, R. A., Harris, D. L., and Green, D. E., *Arch. Biochem. Biophys.* **128**, 219 (1968a): on the effect of *Bordetella* endotoxin upon energized processes in mitochondria.

Harris, R. A., Penniston, J. T., Asai, J., and Green, D. E., *Proc. Natl. Acad. Sci. U.S.* **59**, 830 (1968b): on the correlation of conformational change and functional states.

Jolly, W., Harris, R. A., Asai, J., Lenaz, G., and Green, D. E., *Arch. Biochem. Biophys.* **130**, 191 (1968): on the ultrastructural dislocation induced by lyophilization and the mechanism of uncoupling.

Kopaczyk, K., Asai, J., Allmann, D. W., Oda, T., and Green, D. E., *Arch. Biochem. Biophys.* **123**, 602 (1968a): on the identification of the rutamycin-sensitive ATPase as the headpiece-stalk sector of the repeating unit of the inner mitochondrial membrane.

Kopaczyk, K., Asai, J., and Green, D. E., *Arch. Biochem. Biophys.* **126**, 358 (1968b): on the reconstitution of the tripartite unit from the component sectors.

Krampitz, L. O., Suzuki, I., and Greull, G., *Brookhaven Symp. Biol.* **15**, 282 (1962): on the enzymic interaction of ketoacids with diphosphothiamine.

Kreil, G., and Boyer, P. D., *Biochem. Biophys. Res. Commun.* **16**, 551 (1964): role of phosphohistidine in substrate level phosphorylation.

Lardy, H. A., Graven, S. N., and Estrada-O, S., *Federation Proc.* **26**, 1355 (1967): specific induction and inhibition of cation and anion transport in mitochondria.

Lee, C., and Ernster, L., *Biochem. Biophys. Res. Commun.* **23**, 176 (1966a): competition between oxidative phosphorylation and energy-linked pyridine nucleotide transhydrogenation in submitochondrial particles.

Lee, C., and Ernster, L., *in* "Regulation of Metabolic Processes in Mitochondria" (J. M. Tager, S. Papa, E. Quagliariello, and E. C. Slater, eds.), Elsevier Publishing Co., Amsterdam, 1966b: on the energy-linked nicotinamide nucleotide transhydrogenase reaction.

Lenaz, G., Jolly, W., and Green, D. E., *Arch. Biochem. Biophys.* **126**, 67 (1968): reconstitution of oxidative phosphorylation in lyophilized mitochondria by sonic irradiation.

MacLennan, D. H., and Asai, J., *Biochem. Biophys. Res. Commun.* **33**, 441 (1968): on the identification of the stalk with the protein that determines oligomycin sensitivity.

MacLennan, D. H., and Tzagoloff, A., *J. Biol. Chem.* **241**, 1933 (1966).

Margolis, S. A., Lenaz, G., and Baum, H., *Arch. Biochem. Biophys.* **118**, 224 (1967): stoichiometric aspects of the uncoupling of oxidative phosphorylation by carbonyl cyanide phenylhydrazones.

Mitchell, R. A., Butler, L. G., and Boyer, P. D., *Biochem. Biophys. Res. Commun.* **16**, 545 (1964): role of phosphohistidine in substrate level phosphorylation.

Packer, L., Deamer, D. W., and Crofts, A. R., *Brookhaven Symp. Biol.* **19**, 281 (1966): conformational changes in chloroplasts.

Penniston, J. T., Harris, R. A., Asai, J., and Green, D. E., *Proc. Natl. Acad. Sci. U.S.* **59**, 624 (1968): on the geometric factors in conformational change in mitochondria.

Pressman, B. C., Harris, E. J., Jagger, W. S., and Johnson, J. H., *Proc. Natl. Acad. Sci. U.S.* **58**, 1949 (1967): antibiotic-mediated transport of alkali ions across lipid barriers.

Racker, E., *Federation Proc.* **26**, 1335 (1967): on the mitochondrial ATPase complex.

Reed, L. J., *Vitamins Hormones* **20**, 1 (1962): biochemistry of lipoic acid.
Sanadi, D. R., Gibson, D. M., Ayengar, P., and Jacob, M., *J. Biol. Chem.* **218**, 505 (1956): on substrate level phosphorylation.
Tzagoloff, A., Byington, K. H., and MacLennan, D. H., *J. Biol. Chem.* **243**, 2405 (1968): on the properties of an oligomycin-sensitive ATPase isolated from beef heart mitochondria.

KEY REFERENCES FOR CHAPTER VII

Configurational changes during the energy cycle
Green *et al.* (1968), Harris *et al.* (1968b), Packer *et al.* (1966), Penniston *et al.* (1968)

Changes in configurational mode of the cristae
Hackenbrock (1966), Harris *et al.* (1968a), Penniston *et al.* (1968)

Generation and discharge of the energized state
Green *et al.* (1968), Harris *et al.* (1968b)

Speed of configurational and conformational change
Penniston and Harris (unpublished)

Energized transhydrogenation
Lee and Ernster (1966a,b)

Facilitated entry of metabolites
Griffiths and Haslam (1967)

ATPase complex
Kopaczyk *et al.* (1968a,b), MacLennan and Asai (1968), Racker (1967), Tzagoloff *et al.* (1967)

Uncouplers
Jolly *et al.* (1968), Lenaz *et al.* (1968), Margolis *et al.* (1967)

Respiratory control
MacLennan and Tzagoloff (1966)

Transmembrane delivery of phosphoryl groups
Allmann *et al.* (1967), Beyer (1968)

Antibiotics and other facilitators of salt-induced swelling
Blondin *et al.* (1968), Brierley *et al.* (1966), Lardy *et al.* (1967), Pressman *et al.* (1967)

Configurational changes during swelling
Asai *et al.* (1969)

Energized translocation of divalent metal ions
Brierley *et al.* (1962)

Substrate level phosphorylation
Krampitz *et al.* (1962), Kreil and Boyer (1964), Mitchell *et al.* (1964), Reed (1962), Sanadi *et al.* (1956)

CHAPTER VIII

OTHER
TRANSDUCING
SYSTEMS

SOME INTRODUCTORY CONSIDERATIONS

PRIMARY VERSUS POISED ENERGY TRANSFORMATIONS

Living cells catalyze a wide variety of energy transformations (see Table I.1). It is useful to distinguish between energy transformations in which the input energy is either conserved or utilized for doing work (primary energy transformations) and those energy transformations in which the input energy merely triggers the release of potential energy already accumulated by an independent mechanism (poised energy transformations). The mitochondrion, chloroplast, and acto-myosin network are examples of primary energy transducers; the outer segments of retinal rods and the sensory nerve membrane are examples of poised energy transducers. The primary systems either are actuated by ATP or have the capability to produce ATP; the poised systems neither are actuated by ATP nor produce ATP. The poised systems are energized by virtue of other, primary energy transformations which have to accomplish the preliminary "cocking" of the system. Thus, the latent period in the recovery of the nerve cell or of the retinal rods is an expression of this ancillary independent transduction.

UNIVERSAL PRINCIPLES OF ENERGY TRANSDUCTION

All the energy transformations carried out by the membrane systems* of living cells are reducible to a set of universal principles.

---

* For reasons of simplicity we shall consider the contractile systems as functional equivalents of structured membrane systems.

First and foremost is the concept of conformational change as the molecular basis of energy transformation. In the transformation of energy from the input to the output form, a conformational rearrangement of the system provides the mechanistic link between the two forms. Thus, in the conversion by muscle of the bond energy of ATP into the mechanical work of shortening, the system undergoes profound conformational changes induced by ATP,* and these changes are translated into a shortening of the distance between two points in the system. In the conversion by chlorophyll of the electromagnetic energy of light into the energy of an electron with a high reducing potential, there is a transition involving an excited state of the transducing molecule. The light-induced formation of the "coupled" excited state in chlorophyll† may be formally equated with the ATP-induced conformational change in actomyosin. In the conversion by visual purple (rhodopsin) of the electromagnetic energy of light into an appropriate impulse, there is a conversion of 11-*cis*- to all-*trans*-retinene (retinal) and a concomitant conformational rearrangement of the lipoprotein (opsin) to which *cis*-retinene is bonded. These conformational changes in the components of rhodopsin involve many unstable intermediate states, the end result of the transformation being the release of retinene from opsin with the exposure of a sulfhydryl group in the latter. It is clear from the examples cited that the transitional state in the energy transformation (i.e. the conformationally modified state) may have an intrinsic stability which permits its utilization as such, as in the mitochondrion, or may be highly transitory, as in the outer segments of the retinal rod (Figure VIII.1).

In the primary energy transforming systems, the work performances are carried out by the system in its energized state. In other words, it is conformational energy which drives the various work capabilities of a given transducing system. In the mitochondrion the energized repeating units can synthesize ATP, reverse electron transfer, drive the transfer of hydride ions from DPNH to TPN^+, and concentrate divalent or monovalent cations. In the chloroplast, the energized

---

* In muscle, it is the coupled hydrolysis of ATP and not ATP per se which induces the conformational change.

† Namely, that state which does not degenerate by simple fluorescence.

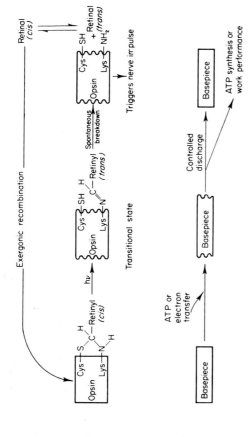

Fig. VIII.1. Comparison between a "poised energy transformation" (the rhodopsin system, upper diagram) and a "primary energy transformation" (mitochondrial inner membrane, lower diagram). The conformationally modified transition state of the rhodopsin system has no intrinsic stability and cannot be utilized to perform a controlled work function.

repeating units* can synthesize ATP or concentrate ions. The consistent rule appears to be that the transducing system is first promoted to its conformationally energized state which is then capable of carrying out the energy transformation intrinsic to the work performance.

Biological energy transducing systems may depend upon a single molecule (rhodopsin), arrays of molecules (chlorophyll), or a complex of several different proteins (the actomyosin complex and the complexes of either the electron transfer chain or the photosynthetic chain) as the instrument for transduction. Within the framework of such variation in the instrumentality of transduction it is permissible to consider energy transformation in biological systems as fundamentally a molecular process. A simple monomeric molecule, or arrays of molecules, or a multiprotein complex is the actual molecular unit of transduction; and, it is the fundamental microprocess in each such transducing unit which, when magnified by large numbers of repeating units arranged in series and parallel, accounts for the macrotransduction.

ATYPICALITY OF THE CONTRACTILE SYSTEM IN RESPECT TO THE MEMBRANE MODALITY

The contractile system, wherever it is found, is the only known exception to the rule that transducing systems are intrinsic parts of membranes. When we examine the contractile system closely we can appreciate why the logistics of muscular contraction have enforced a different modality of molecularization of the transducing assemblies. To achieve symmetrical contraction of muscle in bulk, it is necessary to slide one set of fibers over another, interdigitating, set of fibers, the relative movement being along the axis of the bundle. Each fiber consists of a linear polymeric array of myosin or actin subunits. The interdigitating fibers are arranged in a bundle with a well defined cross-sectional pattern. Each myosin fiber is surrounded by six actin fibers, and each actin fiber is near three myosin fibers. The one-dimensional (linear) modality of polymerization is essential in order to allow each myosin fiber to interact with, and move relatively to,

---

* We refer here to the repeating units of the chloroplast electron transfer chain(s).

all the fibers which surround it (Figure VIII.2). Two-dimensional polymerization would give rise to sliding, interdigitated sheets, asymmetric about the axis of movement, with only two rather than four surfaces of each transducing element (considered as being

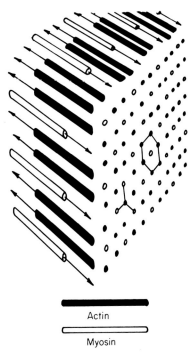

Actin

Myosin

Fig. VIII.2. A diagrammatic representation of the relative movement of actin to myosin. Note the hexagonal spacing, as indicated in the cross section, of actin about myosin, and note that each actin fiber is adjacent to three myosin fibers.

cuboidal) involved in the translation process. The linear arrangement is clearly more desirable in this special case. Theoretically phospholipid could be employed as a device to restrict binding modalities in two dimensions and give rise to linear arrays of repeating units. However, phospholipid is only one of several devices which can restrict interactions between protein subunits. Its effect can be duplicated by other charged groups which are part of the protein itself and localized on the appropriate face(s) of the repeating unit. In the case of the contractile system it is, indeed, charged groups on the protein subunits which determine the mode of polymerization. These are probably the very groups that are involved in the transduction process itself.

INTRODUCTION TO SOME OF THE TRANSDUCING SYSTEMS OF THE CELL

There are some dozen different types of membrane systems which implement the most essential transductive processes of living systems. In bacteria all transductive systems appear to be associated with one mammoth membrane* whereas in animal and plant cells a given transductive system is localized in any one of a number of separate membranous organelles. This, of course, is not a fundamental distinction in the sense of mechanistic principle. The macromolecular components of the transducing systems of bacteria and the molecular principles of these transductions are probably very similar to the corresponding components and principles that apply to cells in which the membranes are in separate organelles.

We shall attempt in this chapter to introduce the reader to the important categories of transductive events in living cells. A great deal is known about some of these events (muscular contraction, transmission in nerve, photosynthesis) and little or nothing about some of the other events (cell division, information storage). It would be not only impossible but undesirable to attempt within the precincts of a single chapter to describe these various transductive events *in extenso*. Our aim will be the more modest one of specifying, with as much precision as the facts permit, the essential nature of the transductive event and the kind of molecular machinery that is

---

* Tubular spurs from the bacterial membrane may each correspond to an organelle. Recent electron micrographic studies by Remsen and Watson on marine microorganisms have revealed that the electron transfer system is associated with a system of internal membranes that is completely analogous to the cristael membranes of mitochondria. The internal membranes of these bacteria originate from the plasma membrane in contrast to the cristae which originate from the inner boundary membrane of the mitochondrion. But, apart from this one difference, the internal membranes of the marine organism and the cristae are indistinguishable morphologically.

required for implementing the transduction. In so doing we shall from time to time present models of a number of systems. These models are intended as illustrative of the principles which might be involved in the working of such systems. They are not intended to be arbitrary or in any sense to represent value judgments in fields which are beyond our working experience and where rival models might legitimately claim greater validity. Moreover, in presenting such selective descriptions we intend in no way to diminish the contributions made by distinguished scientists whose work we may appear to have overlooked.

CONTRACTILE SYSTEMS

Living cells have evolved a wide variety of systems (skeletal muscle, cardiac muscle, flagella, cilia, trichocysts) for achieving rotational or translational movement of one part of a multicellular organism with respect to the other parts, or of one sector of a cell with respect to other sectors. The hallmark of the contractile system is the speed and power of the movement, whatever form it takes. The molecular instrumentality of virtually all contractile systems is the actomyosin system—a system of at least two interdigitating polymeric molecules, namely myosin and actin. The study of the actomyosin system has proved to be one of the most exciting and fruitful adventures in molecular biology; it provides valuable guideposts both in research tactics and in theoretical principles.

The progress made in the study of muscular contraction stems from the relative chemical simplicity of the system and from the imaginative and powerful use made of the electron microscope. In the case of muscle, the easiest way to study the transductive event (contraction) was to see it happening, and this demanded ultrastructural observations of great sophistication. This technical restraint was a blessing in disguise. In the case of the myofibril, the contractile event could be categorically defined in ultrastructural terms as the sliding of one set of filaments over another (see Figure VIII.3), without (in the first instance at least) any shortening or thickening of the filaments themselves.

The thick filaments are almost certainly aggregates of the protein myosin (which comprises about 40 % of the total muscle protein in vertebrate, striated muscle). The thin filaments are made up of fibers of actin (15 % of the total muscle protein). A great deal is known about the chemistry of these two proteins and about the size and shape of their molecules. Thus, much is known of the way in which myosin molecules can aggregate into fibers or can be broken down into "meromyosin" subunits. Myosin is also known to exhibit a characteristic ATPase activity. Actin can be depolymerized reversibly to form globular units (actin G) and repolymerized into fibrous actin F. Mixtures of actin and myosin form an "actomyosin" complex. When the complex is formed under suitable conditions, actomyosin can be extruded as a thread with mechanical and chemical properties very

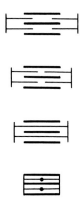

Fig. VIII.3. Sliding of actin fibers over a myosin fiber as the basis of muscular contraction.

similar to those of actomyosin *in vivo* or to those of actomyosin in glycerol-extracted muscle cells. However, the chemical facets had to be rationalized with ultrastructural facets. Thus, the chemical events underlying contraction had to be described in terms which would account for the vectorial movement of the actin filament along the myosin filaments without the simultaneous thickening of either. The possibility arose that the extensively studied ATPase activity of myosin, or the actin F-actin G transformation, or the properties of extruded actomyosin threads might prove to be irrelevant artifacts or misleading manifestations of relevant properties. Guided, therefore, on the one hand by detailed chemical studies on the composition and organization of the transducing elements, and on the other by the unambiguous realities of the ultrastructural events associated with contraction, workers in the field are now developing meaningful, if approximate, descriptions of the chemical events which underlie the contraction process.

The following is a synopsis of one such description: F-actin is an open, two-stranded helix of G-actin subunits, with about 13 units

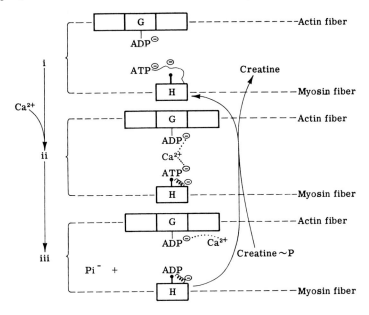

Fig. VIII.4. Diagrammatic representation of a hypothetical series of events involved in the translation of an actin fiber relative to a myosin fiber. G = G-actin subunit; H = H-meromyosin subunit; ✿ ATPase site.

per turn of the helix. To each G-actin subunit is firmly attached a molecule of ADP. Each myosin fiber is a polymer of about 200 myosin molecules, each molecule being tadpole-like in shape, the "head" consisting of a so-called H-meromyosin subunit. During muscle contraction, actin fibers are translated toward each other over each myosin fiber to meet (or overlap) at its center (see Figure VIII.3). The polymeric myosin fiber is "polarized" about this center, the constituent myosin molecules being arranged with their tails toward and their heads away from this point. The mode of poly-merization is such that the heads (H-meromyosin) coat the fiber in an array with a sixfold axis of symmetry (so that each actin fiber of the set which surrounds every myosin fiber is faced with an identical row of H-subunits). From each H-subunit, directed away from the center of the fiber, extends a negatively charged polypeptide chain. ATP binds to the end of this polypeptide which, by virtue of negative charge repulsion, extends as a long snaking chain (Figure VIII.4,i).

Contraction is initiated by the release of Ca^{2+} from the sarcoplasmic reticulum (this release being initiated in turn by a wave of depolariza-tion conducted from the sarcolemma). Each released Ca^{2+} ion forms a strong chelate, bridging an ADP on a G-actin subunit with an ATP on the end of an extended polypeptide chain. This chelation neutralizes the excess negative charge on the chain and allows the chain spontane-ously to assume an α-helical configuration stabilized by hydrogen bonding and hydrophobic interactions. Helix formation shortens the chain and drags the actin fiber along the myosin fiber (Figure VIII.4,ii). The cross link between the two fibers is now broken by the hydrolysis of the bridging ATP molecule at an enzymic site on the H-meromyosin, which has now been brought into apposition with this bridge. This leaves a helically coiled side chain with a molecule of ADP attached (Figure VIII.4,iii). Phosphorylation of this molecule by a catalyzed phosphoryl transfer from external phospho-creatine again introduces an excess of negative charge into the side chain, which then unwinds and is thus recharged in preparation for another cycle of activity (which again results in moving the fibers another "notch" along relative to each other). The myofibril continues to contract in this manner until these cycles of events are halted by the intervention of "relaxing factor." Probably such "factors" represent nothing other than agents for the removal of Ca^{2+}. The key "factors," indeed, is the sarcoplasmic reticulum which actively accumulates Ca^{2+} by a process of energized translocation manifesting itself as a Ca^{2+}-stimulated ATPase (see below).

Molecular models of the kind presented above are no more than inspired interpretations of the available facts and merely serve as aids to visualization. There can be no doubt that ATP energizes the process of muscular contraction, that Ca^{2+} mediates the discharge of the energized state, and that each cycle involving energizing and discharge leads to a movement of the actin fibers relative to the myosin fiber. It is also reasonably certain that there is a cycle of attachment and detachment of the actin and myosin fibers. But, whether there is a transition from random coil to helix of the polypeptide chain of the H-meromyosin unit during the cycle of attachment and detachment is still an open question. Of particular interest to the theory of energy transductions in membrane systems is the fact that ATP induces a series of conformational changes by which one set of fibers can be translated relative to another set.

The contractile system operates in only one direction. ATP can induce "contraction," but extension of the muscle (i.e. the application of an external stretching force) does not lead to synthesis of ATP. This is in no way surprising; the contractile apparatus is not designed to function as a delicately poised reversible system.

The ATPase associated with preparations of actomyosin is inactive in hydrolyzing ATP unless ions of a divalent metal (Mg^{2+} or Ca^{2+}) are present. That requirement is satisfied by Mg^{2+} but not by Ca^{2+} when a minor protein fraction is extracted out of the preparation. When this protein recombines with the extracted actomyosin system, then Ca^{2+} can once again mediate the controlled hydrolysis of ATP. Much the same applies to the mitochondrial ATPase. Ca^{2+} does not activate the ATPase associated with the headpiece-stalk sector of the repeating unit of the inner membrane, but it exhibits its full activating capacity when the headpiece-stalk sector is functionally coupled to the basepiece. Under such coupling conditions, calcium-activated ATPase activity leads to translocation of Ca^{2+} and P_i.

The actomyosin system is embedded in the structured matrix of the muscle cell. The regular arrays of actomyosin fibers are interdigitated with regular arrays of mitochondria and tubules of sarcoplasmic reticulum (Figure I.1). It is the mitochondrion which generates ATP in the immediate neighbourhood of a set of actomyosin fibers, and it is the sarcoplasmic reticulum which controls the delivery and sequestering of Ca^{2+} for that same set of fibers. Thus, there are sets of organelles in repeating domains of the muscle cell, which collaborate in the implementation of contraction and relaxation. In some types of muscle cells of the primarily anaerobic variety, it is the glycolytic system rather than the mitochondrion that generates the ATP.

THE CHLOROPLAST AND THE PHOTOSYNTHETIC APPARATUS

The chloroplast (Figure VIII.5) is probably the most complex of all biological machines. Its input is the primary biological "fuel," solar radiant energy, together with CO_2 and water. Its output is oxygen plus certain organic molecules, such as sugars, which serve

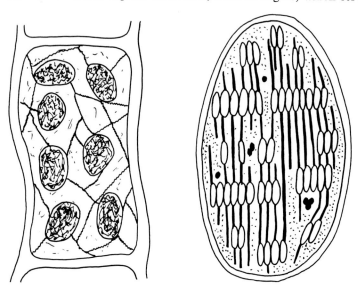

Fig. VIII.5. A diagrammatic representation of chloroplasts, as seen in brown algae (at left) and of a single chloroplast from barley (at right).

as the secondary fuels for other living cells. Thus, the chloroplast catalyzes the elaborate process summarized by the equation:

$$6CO_2 + 6H_2O + h\nu \rightarrow C_6H_{12}O_6 + 6O_2$$

This process can, however, be broken down into three systems of reactions*:

1. The generation of net reducing power by the photolysis of water,

$$H_2O + 2h\nu \rightarrow 0.5O_2 + 2H^+ + 2e^-,$$

with the concomitant synthesis of ATP;

* A similar set of collaborating systems is found in phototropic microorganisms.

2. The generation of ATP by coupled phosphorylation.

3. The utilization of the reducing power and the ATP to drive the reductive synthesis of carbohydrate from CO_2.

Clearly, the third system carries out a set of reactions which formally may be considered equivalent to the reversal of the Krebs cycle. In the latter case, pyruvate (derived from carbohydrate) is converted to CO_2 and reducing equivalents, and ATP is formed at the substrate level. In the case of photosynthesis, reducing equivalents are used to synthesize sugars from CO_2, and ATP is required to drive the reaction "uphill." The sequence of chemical reactions involved in the reductive synthesis of sugar in chloroplasts is different from the oxidative dissimilation of pyruvate in the mitochondrion, but undoubtedly the same principles apply.

The two transducing systems, Systems 1 and 2, will be our immediate concern. They are certainly very complex and are, moreover, integrated with one another. There are many models which attempt to describe the mechanism of these processes, but it is not within the scope of this book to attempt to evaluate them. Instead, we shall construct a simplified model system which points out the several events involved and their close analogy with the process of oxidative phosphorylation.

All photosynthetic systems are membranous, but there is considerable variation in the ultrastructural appearance of the photosynthetic membranes. In bacteria such as *Chromatium*, the photosynthetic system is localized in vesicular membranes with little regularity. In higher plants, the photosynthetic system is contained in an organelle, the chloroplast, which is generally somewhat larger than the mitochondrion (see Figure VIII.6). The chloroplast is made up of three membranes—the outer membrane enclosing the organelle, and an inner membrane system differentiated into lamellae and granae. The inner chloroplast membrane appears to be a composite of two interdigitating membranes—the granae and the lamellae. The granae are the folded, stacked membranes which interrupt periodically the lamellae that otherwise extend continuously from one end of the chloroplast to the other. The granae have been reported to be missing in mutants which lack the capacity to evolve oxygen during the photosynthesis. If, indeed, the oxygen-evolving photosynthetic system is localized in the granae, the relative simplicity of the structure

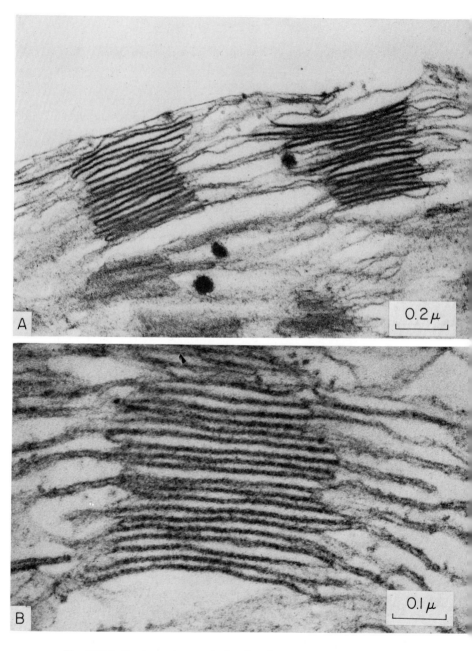

Fig. VIII.6. Electron micrograph showing the granae of spinach chloroplast as tubular membranes in the center of a set of overlapping lamellae membranes. The magnification of (B) is twice that of (A).

of photosynthetic systems which lack the oxygen-evolving capability could readily be rationalized. These systems would have only the lamella type of membrane.

The principal molecular instrument for gathering the electromagnetic energy of light is chlorophyll. (Carotenoids are also concerned in this process but more in a capacity subsidiary to that of chlorophyll.) Two features of chlorophyll are diagnostic of this tetrapyrrolic structure—the presence of Mg as the central metal atom, and the presence of a phytyl chain linked to a propionic acid residue on the ring IV (see Appendix II). There are two principal kinds of chlorophyll—a and b—which differ only in the nature of the substituent in ring II. The absorption spectra of the two chlorophylls cover a wide swath of the range of wavelengths characteristic of visible and near infrared light.

In photosynthetic systems, chlorophyll is found in two forms—as free chlorophyll and as protein-bound chlorophyll. Free chlorophyll is some 50–100 times more abundant than protein-bound chlorophyll. It is a reasonable inference from the available experimental evidence that the unit of photochemical action is lipoprotein in nature and corresponds to a repeating unit of the membranes which make up the photosynthetic system. Each such repeating unit contains six molecules of protein-bound chlorophyll and at least 300 molecules of free chlorophyll (a or b). The free chlorophyll appears to be localized in the lipid phase of the repeating unit in a highly oriented fashion. The molar ratio of phospholipid to chlorophyll in some active photosynthetic particles has been found to be 1:1.

The chlorophyll-protein compound is often referred to as P_{770}, P_{830}, etc., depending on the absorption maximum of its principal light-absorbing band in the near infrared region. Since the value of the absorption maximum for this compound varies widely from one photosynthetic system to another, we shall refer to this protein as C-P. Within each lipoprotein repeating unit, C-P is associated with a set of molecules (pteridin, cytochrome, etc.) which are essential for the primary photosynthetic event. The ensemble of C-P and associated oxidation-reduction molecules, and of lipid, carotenoid, and free chlorophyll, constitutes one operational unit as well as the morphological unit. The operational unit is the light-gathering center and the photochemical reaction center. The morphological unit is the repeating unit which makes up the membrane continuum. We may define the photochemical reaction center as the system which contains C-P in its proper geometric relation to all the other reactants in the repeating unit which participate in the primary photochemical process. C-P has been isolated from green bacteria and shown to have a molecular weight of 137,000 and to contain 6 molecules of chlorophyll per unit of molecular weight.

The repeating units of the photosynthetic membranes are probably tripartite in nature. Moudrianakis has isolated from spinach chloroplasts an ATPase which dimensionally corresponds to a headpiece. In our laboratory, headpiece-stalk sectors have been clearly demonstrated by Asai *et al.* in negatively stained specimens of chloroplasts. Furthermore, Parks has shown that the quantasome, which corresponds to the functional and morphological unit of photosynthesis, is associated with the membrane-forming sectors of the chloroplast membrane. From these observations we may provisionally conclude that all the repeating units of the photosynthetic membranes (inner member of the chloroplast) have tripartite repeating units, that the headpiece of these units is the site for coupled phosphorylation and the basepiece the site for any of the several photosynthetic processes (the primary photosynthetic event or one of the subsequent electron transfer processes). It should be pointed out that all repeating units in the inner mitochondrial membrane are tripartite in nature, not only the ones concerned in coupled phosphorylation.

When light impinges on the light-gathering centers, a photon is absorbed by one of the free chlorophyll molecules in the closely packed array of several hundred molecules present in the center. The excited state produced in this acceptor molecule is transmitted by some kind of resonance phenomenon through the array of chlorophyll molecules until the excitation reaches the chlorophyll molecules intrinsic to C-P. All of this transfer takes place without significant loss of energy. C-P in the excited state catalyzes an "uphill" transfer of electrons from an associated electron donor (probably a *c* type cytochrome in its reduced form) and an associated electron acceptor (pteridine). This is merely a balance sheet account of the photochemical transaction, not an explanation. The essential point is that C-P in the excited state can catalyze the transfer of an electron from a reduced cytochrome of the *c* type to pteridine—which as a group has unusually negative oxidation-reduction potentials (about -0.7 V). The light-energized uphill electron transfer from cytochrome to pteridine generates an electromotive force of about 1 V—an electromotive force sufficient for generating at least 3 molecules of ATP by coupled phosphorylation (see Appendix I).

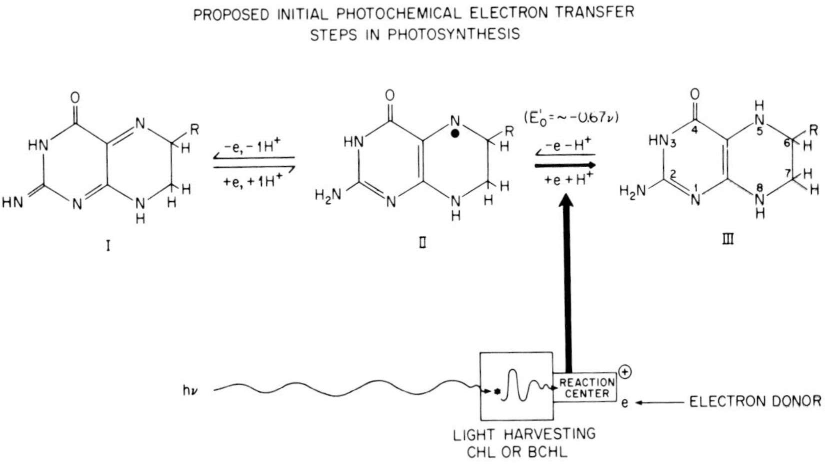

PROPOSED INITIAL PHOTOCHEMICAL ELECTRON TRANSFER
STEPS IN PHOTOSYNTHESIS

Fig. VIII.7. A proposed formulation of the initial photochemical electron transfer step in photosynthesis. I = dihydropteridine; II = pteridine semiquinone; III = tetrahydropteridine. I is 2-amino-4-hydroxy-6-substituted pteridine. CHL = Chlorophyll; BCHL = Bacteriochlorophyll. Taken from R. C. Fuller and N. A. Nugent, *Proc. Natl. Acad. Sci. U.S.,* August issue (1969).

The reduction of pteridine can proceed in two one electron steps (see Figure VIII.7). Since one electron can be transferred to pteridine simultaneous with each photochemical event, the absorption of one photon by chlorophyll *a* leads to the transfer of one electron to a pteridine molecule in the reaction center.

Fuller and Nugent who were the first to recognize the role of pteridine in photosynthesis have provided evidence that pteridine is an absolute requirement for photosynthesis in green bacteria. The photochemical reaction center would appear to be specific for pteridine as the photooxidant. Whether this specificity also applies to the photoreductant is still an open question.

Considerable progress has been made in defining the components of the electron transfer chain of photosynthetic bacteria. In a general way, the electron transfer chain is essentially identical with that of the mitochondrion except in two respects. First, the ultimate reductant is pteridine and not succinate or DPNH. Second, the terminal end of the mitochondrial chain (Complex IV) is eliminated in the photosynthetic chain. The probable sequence of electron flow in the photosynthetic electron transfer chain may be formulated as follows: pteridine $\rightarrow$ ferredoxin $\rightarrow$ flavoprotein (DPN$^+$) $\rightarrow$ coenzyme Q $\rightarrow$ cytochromes b and c_1 $\rightarrow$ cytochrome c. At least one complex of the photosynthetic electron transfer chain can be defined, namely the complex which is the equivalent of Complex III of the mitochondrial electron transfer chain. This identification is justified on the basis of the sensitivity of the photosynthetic chain to antimycin—an inhibitor highly specific for Complex III. This complex which contains cytochromes b and c, and non-heme iron would catalyze the oxidation of reduced coenzyme Q by cytochrome c (or its equivalent in the photosynthetic bacterial cell). It would appear that there are at least two other complexes, both containing flavoprotein—one which catalyzes the oxidation of reduced ferredoxin by coenzyme Q and another which catalyzes the oxidation of DPNH by coenzyme Q. The presence of two complexes, both of which interact with coenzyme Q, makes possible the generation of DPNH as a side path in the electron transfer process (the main pathway presumably does not involve the oxidoreduction of DPN$^+$). DPNH or TPNH (these are always in equilibrium) is the ultimate reductant of CO_2 in carbohydrate synthesis.

On the basis of two complexes in the photosynthetic electron transfer chain covering the range of potential from that of ferredoxin at the reducing end to that of cytochrome c at the oxidizing end, only two sites for coupled phosphorylation would be available. Thus far, only two sites have indeed been found. To achieve a higher yield of ATP in photosynthetic phosphorylation, there would have to be a complex spanning the potential drop from reduced pteridine to ferredoxin. There is no available evidence for such a possibility. Presumably the energy drop between reduced pteridine and ferredoxin is wasted in the photosynthetic chain just as the energy drop between cytochrome a and oxygen is wasted in the mitochondrial chain.

All the available evidence is compatible with the assumption that the mechanism of photosynthetic phosphorylation is identical with that of oxidative phosphorylation in mitochondria. Figure VIII.8 shows the profound configurational changes in isolated spinach chloroplasts during the dark-light transition. These configurational changes are suppressed by uncouplers such as m-chlorocarbonyl-cyanide phenylhydrazone.

The chloroplast like the mitochondrion can link electron flow not only to the synthesis of ATP but also to ion translocation. This correspondence in coupling capability is yet another token of the mechanistic identity of the coupling process in the two organelles.

There is in chloroplasts a separate photochemical system which catalyzes the transfer of electrons from water to some photooxidant, possibly coenzyme Q. This transfer leads to the evolution of molecular oxygen and the reduction of the photooxidant. We may postulate a photochemical reaction center, built into a lipoprotein repeating unit, which contains C-P, the light-gathering accessories (arrays of free chlorophyll and carotenoid), and specialized molecules which determine the electron transfer from water to the photooxidant. Chlorophyll in principle can undergo either photoreduction or photooxidation. Which will take place will be determined by the special conditions and environment of C-P in the context of the photochemical reaction center.

Since the oxygen-evolving photochemical system is part of the same membrane network which contains the electron transfer chain, it follows that the photochemical evolution of oxygen will also be accompanied by coupled phosphorylation. Reduced coenzyme Q generated during this process will be oxidized via the electron transfer chain, and this oxidation will lead to ATP synthesis. Such a tie-up of the oxygen-evolving system and the electron transfer chain is only possible with another tie-up between the electron transfer chain and the pteridine-dependent photochemical system. Without this tie-up, the oxidation of reduced coenzyme Q would stop as soon as the terminal oxidant, cytochrome c, became reduced. Regeneration of cytochrome c in the oxidized form requires the operation of the pteridine-dependent photochemical system; it also requires that as many electrons leave the system via the DPNH or TPNH side path as enter it from the photolysis of water.

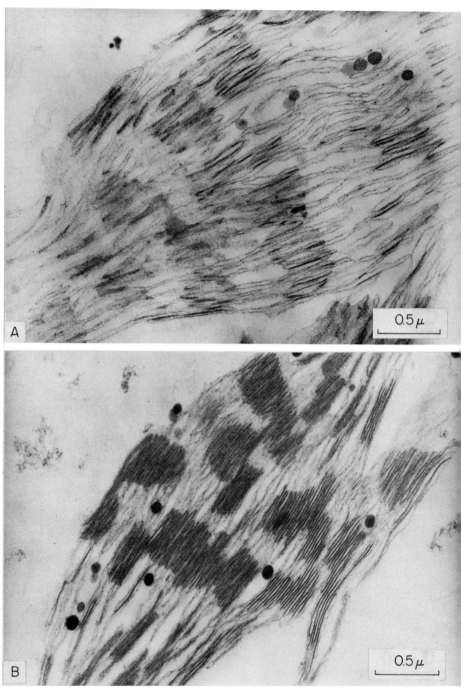

Fig. VIII.8. Configurational changes in the granae of spinach chloroplasts induced by light. A: in the dark; B: in the light.

MEMBRANES WITH TRANSPORT CAPABILITY

All membranes that delimit cells (plasma membranes), as well as some of the membranes in the interior of a cell, have a capability for transporting particular molecules or ions. The outer membranes of some organelles, such as the mitochondrion, are the only membranes known to be without the capability for active transport. This capability has been correlated with the presence of tripartite repeating units (macro or micro) in the membrane. The list of transportable molecules and ions covers the range from inorganic salts to macromolecules and from substrates of enzymes to enzymes. By membrane transport we mean to imply that the passage of a molecule or ion through a membrane involves more than random diffusion and that there is a mechanism or device by which the passage through the membrane is facilitated. There are two kinds of transport—energized and nonenergized. In energized transport, the direction of movement of molecules and the extent of the movement are not governed merely by the gradient established. Energy is utilized to move molecules unidirectionally and to build up concentrations in the cell or membrane which are higher than the concentrations outside the cell or membrane. In nonenergized or facilitated transport, there is no driving force other than the difference in chemical potential of the molecules external and internal to the membrane barrier. Independent of the kind of transport some special mechanism has to be available for permitting a particular molecule to penetrate the membrane barrier. Without such a mechanism, there is no penetration of the membrane (apart from slow diffusion of small molecules between repeating units).

Whether the transport is energized or facilitated, it would appear that the membrane contains a protein (permease) which can bind specifically to the molecule which is to be transported. Binding to a specific protein in the membrane appears to be a *sine qua non* for transport of different kinds of molecules and ions (lysine, methionine, sulfate, K^+, glucose, lactose). How the binding to the permease facilitates the next step in the transport is not clear. In energized transport a conformational change in the membrane may make it possible for the binding protein to fluctuate in position from one side of the membrane to the other. But what would be the mechanism for facilitated transport?

A massive research effort is now being directed toward the demonstration of isolatable permeases in membrane systems and toward the correlation of these permeases with the transport capabilities of the membranes in which these are localized. If the mechanism by which ions are transported across the mitochondrial membranes is valid for all membrane systems (and the principle of biochemical universality would support such an extrapolation), then it would follow that all active movements of ions and molecules through a membrane depend on the same type of cyclical mechanism of generation and discharge of the energized state. In other words, there may well be a general, energized transport mechanism independent of the nature of the molecules transported. This mechanism has to do with the conformational basis of the energized state of membrane systems. Specificity is grafted onto the general translocating system by permeases which stand in the same relation to transportable molecules or ions as does an enzyme to its substrate. It is possible that the permease is a subunit of a basepiece of a membrane and so disposed within the basepiece that during the cycle of generation and discharge of the energized state, the permease is alternatively first on one side and then on the other side of the membrane. The fact that all membranes which carry on active transport have tripartite repeating units would suggest that the structure of the repeating unit is relevant to the universal mechanism of active transport referred to above.

MEMBRANE SYSTEMS WITH THE CAPABILITY FOR ENERGIZED TRANSPORT

This brings us to a consideration of the problem of the active transport of specific solutes across specific biological membranes against concentration gradients. The sarcoplasmic reticulum catalyzes the ATP-driven transport of Ca^{2+}. The erythrocyte plasma membrane catalyzes the ATP-dependent exchange of Na^+ for K^+. The membrane potential of nerve axons (the "charge" which pays for the propagation of nerve impulses, however triggered) is maintained by a similar active transport process. Functions of the kidney tubular epithelial cells*

* The proximal limb of the kidney nephron is a long winding tubule of epithelial cells whose boundaries are so dovetailed that the lumen could be considered as one continuous membranous surface. This membrane exhibits different transport capabilities at different sections of the tubule. We must envision the likelihood that the repeating units change in composition from one end of the tubule to the other. This change in composition reflects the different binding proteins or permeases required for transport of molecules or ions.

and the acid-secreting parietal cells of the gastric epitheleum may also be defined in terms of active transport of specific solutes. The solutes translocated across these membranes may be inorganic ions or organic molecules such as glucose or amino acids. The transfer may be into the cell (e.g. uptake of glucose) or out of the cell (e.g. secretion of H^+). Moreover, in the case of ion translocation the end result may be manifest in such different ways as the production of a concentration gradient (e.g. HCl secretion) or an electrical potential (e.g. nerve).

In spite of their variety, all these membrane-centered processes have certain characteristics in common. Thus, each translocase apparatus is specific for a single "substrate" solute or at least for a limited class of such solutes. Moreover, there is evidence for the existence in translocating membranes of proteins (the permeases referred to above) with specific binding capabilities for such "substrate" solutes. Further analogies with simple enzymic systems are suggested by the facts that active transport exhibits saturation kinetics and that there are inhibitors (both of the competitive and of the noncompetitive types) which are known to be specific for certain translocation processes only.

In addition to their similarities to each other, all active transport processes have the characteristics of primary transducers of energy. They are membrane-centered, molecular devices which perform work functions energized by ATP. Moreover, the very nature of the work performance in question, the vectorial movement of solute molecules against a chemical potential gradient, implies the intervention of a carrier capable of participating in the energized conformational change.

The analogy is particularly marked between certain cation translocating systems of the plasma membrane, on the one hand, and the ATP-energized accumulation of cations by the mitochondrion on the other. Among the systems in the first group is the one which transports Na^+ ions (outward) in exchange for K^+ ions (moved inward) at the expense of the hydrolysis of ATP. This ATPase activity* is stimulated by K^+ ions on the outside of the membrane

---

* The ATPase of several membranes specialized for transport of K^+ into the interior and transport of Na^+ out of the interior has been shown to involve the formation of a phosphoryl enzyme. We may consider the phosphoryl enzyme as the "form" of ATP which initiates conformational change in the membrane, rather than ATP itself.

and by Na^+ ions on the inside; when the location of these ions is reversed, the hydrolytic activity is inhibited. This inhibition, when polarity of localization of the ions is reversed, may be one of the control features of the membrane. The cardiac glycoside, ouabain, specifically and strongly inhibits this ion transporting system and the ATPase activity which is coupled to the transport. In this respect ouabain acts in a manner analogous to oligomycin (rutamycin), which prevents the generation by ATP of the energized conformational state of the repeating units of the mitochondrial inner membrane. As in the case of the mitochondrial system, the Na^+-, K^+-ATPase system of the plasma membrane loses most of its native characteristics when the membrane is dispersed by detergents. When the detergents are removed and the repeating units are allowed to revesicularize, the ouabain-sensitive, Na^+-, K^+-dependent ATPase activity reappears.

THE SENSING SYSTEM

Multicellular organisms, and to a more limited extent unicellular organisms, contain specialized systems which are uniquely triggered by any one of several types of stimuli (radiant and sound energy, temperature gradients, pressure, chemical reagents). When triggered by the appropriate stimulus, these membranous systems respond in a characteristic way. The response in multicellular organisms usually takes the form of a perturbation that eventually initiates a nerve impulse to the central nervous system. The sensing systems have to be cocked by a series of enzymic events that precede the triggering. Only in the case of the photoreceptor system have we information about the nature of the reactions involved in the cocking mechanism. It is highly likely that each sensing system will have its own unique cocking mechanism. By the same token we may expect that the macromolecule that responds to light will be different from the molecule that responds to sound or to a chemical stimulus. However, it is reasonable to expect that the perturbation of the sensing systems will involve much the same *kind* of response (independent of the nature of the stimulus that triggers the perturbation) because sensing systems eventually must interact with nervous elements and this interaction in common would call for a common or similar mechanism for initiating the nerve impulse.

The photoreceptor molecule in the repeating units of the inner membrane of the outer segments of the retinal rods is (as we have already mentioned) a protein bonded to 11-*cis*-retinene (see Figure VIII.1). In the cocked state of the membrane, a certain proportion of the photoreceptor molecules are intact, i.e. the carotenoid moiety is predominantly in the 11-*cis* form and conjugated with opsin. When light strikes some of the photoreceptors in the organelle, the carotenoid moieties are transformed into the all-*trans* form; they then dissociate from the protein to which they are conjugated. The conformational changes involved in the *cis-trans* transition and in the dissociation of the chromophore from its apoprotein reveal a free sulfhydryl group in the latter (see above) and lead to an extensive rearrangement of the membrane and a movement of protons.

When the triggering by light is completed, the carotenoid moiety must be transformed back to the 11-*cis* form from the all-*trans* form (which then combines with opsin again to regenerate the photoreceptor). An elaborate series of reactions is involved in this repair process (which probably takes place elsewhere in the retina), and the repair process is, thus, necessarily slow compared to the speed of the triggering by light. It would appear that only a fraction of the total number of photoreceptor molecules in the receptor elements is affected by light at any one time, since not all the photoreceptors are in the cocked state.

The retinal rods are organelles with outer limiting membranes enclosing internal stacked discs (Figure VIII.9). We are still uncertain of the locale of the enzymes which implement the transformation of retinal photoreceptors from the all-*trans* to the 11-*cis* form. In the dark-to-light transition, the discs *in situ* undergo a marked change in shape (from expanded to flattened tubules), but there are technical difficulties in demonstrating these changes consistently with isolated retinal rods.

The outer segment of the retinal rod is attached to a base structure, the ellipsoid (inner segment), via a contractile cilium (Figure VIII.9). The ellipsoid contains sets of closely nesting mitochondria as well as the nucleus. When light induces contraction of the discs of the retinal rods, there is often seen an accompanying change in conformation of the cristae of the mitochondria from the expanded energized state to the contracted nonenergized state. Similarly, the actomyosin system

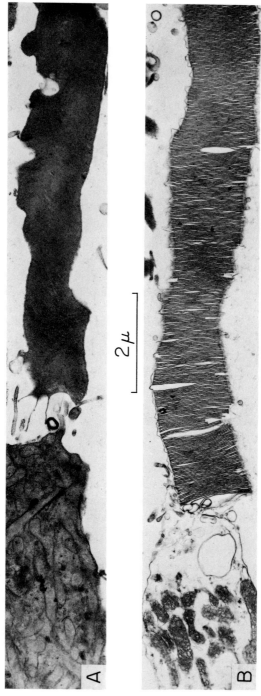

Fig. VIII.9. Electron micrograph showing the outer segments of the rods of beef retina as well as the basement membranes. A: in the light; B: in the dark.

of the cilium undergoes a transition from the resting to the contracted state. Thus, light, by initiating a conformational rearrangement of receptor molecules in the retinal discs, triggers a response in all the organelles associated with a bank of discs. Nothing is yet known of the nature of the trigger substance released by light-induced contraction of the retinal discs which compels the mitochondrion to fire; and, in turn, it is not known what compels the cilium to contract. These events may be unrelated to the initiation of the nerve impulse. At any rate, the interdependent system of organelles in a sensory receptor ensemble poses some fascinating problems of membrane-membrane interactions.

The hallmark of a sensory system is its sensitivity to extremely small pulses of input energy, be it light, sound, heat, or chemical binding. The response, in terms of the transmission of a nerve impulse, involves energies which are one or more orders of magnitude larger than the input energies. Clearly, there has to be an amplification of the signal to a point at which a nerve can be fired. How is this amplification achieved? The retinal discs have to be cocked by a set of dark reactions which convert all-*trans* retinal into the 11-*cis* form. This is, however, only part of the energy put into the system. When the discs are fired by impinging photons, it is probable that only a small proportion of the total number of repeating units is directly fired by light. The other repeating units are discharged in the wave of discharge which presumably attends the deenergizing of a critical number of repeating units in a membrane. This would constitute the first amplification factor. The second factor, and probably the most important, is the collaboration between the retinal discs and the associated set of mitochondria.

SYSTEMS FOR NERVOUS EXCITATION

In multicellular organisms there is a network of nerve cells that transmit an excitation (i.e. an electrical impulse or "action potential") from the peripheral endings of these nerve cells to some receiving center. The transmission is accomplished by the membrane that encloses the cell and its axon. This membrane is cocked by an elaborate process as yet not fully defined. During this cocking process an electrical potential is established across the membrane, i.e. between its inner surface and the external medium. This potential difference is referable to the functioning of an ion pumping system (of the kind that we have previously discussed), which maintains a much higher concentration of K^+ and a much lower concentration of Na^+ within the nerve cell and its axon as compared to the corresponding concentrations in the external medium.

The maintenance of ion gradients by such pumping mechanisms depends in part upon the relative impermeability of the membranes to ion movement. As in the case of the mitochondrion, specific agents, such as gramicidin, greatly increase the permeability of some membranes to certain cations; hence they can discharge ion gradients previously established. In the case of nerve cells, such permeability changes are initiated by the trigger mechanisms of the sensory transducing systems or by neurotransmitters such as acetylcholine. The transient current flowing during such a discharge increases the permeability of an adjacent portion of membrane; this results in a flow of ions, which in turn, disturbs the next portion of the membrane. Hence, there is a propagation of the disturbance along the previously primed nerve axon. As soon as the "action potential" has passed, the ion pumping system resumes the task of recharging the membranes.

Thus, the depolarization process itself is not an energized process mediated or powered by ATP, but merely an induced change in membrane permeability. Whether the spontaneous movement of cations down the electrochemical gradient of the cocked membrane during depolarization involves a true random diffusion or a facilitated transfer via a permease is not known.

The nerve membrane is really a specialized variant of the general class of poised sensing systems. The sensing systems communicate to the nerve membrane the fact of their discharge but the general mechanism by which the sensing systems and the nerve membranes are discharged or depolarized may be very similar indeed. It is the transmission of an excited state of a membrane (conduction of a nerve impulse) that is the essential point of difference between the sensing systems and the nerve membrane. In conduction along a nerve the excited state is transmitted over considerable distances, whereas in the operation of the sensing systems, the disturbance is propagated over relatively short distances (only to the point of interaction with a nerve ending).

RELAY SYSTEMS

The transmission of excitation from one nerve to another along a particular pathway (synaptic transmission) or from a motor nerve to a muscle (neuromuscular transmission) involves the mediation of chemical neurotransmitters, such as acetylcholine and noradrenaline (norepinephrine). These substances are released from membrane-bound synaptic vesicles which lie in the cytoplasm of the nerve axon near its terminal synaptic or neuromusclar junction. The release of transmitter in a controlled packet (one vesicle is probably emptied per impulse) occurs when a wave of polarization sweeps toward this terminal region of the axon. The mechanism of release of transmitter substances is poorly understood, but the release process may be provisionally classified as a variant of the release process in sensing systems. Once released, the transmitter affects the permeability of the adjacent muscle or nerve membrane and initiates a new wave of depolarization.

SYSTEMS FOR THE RECEPTION OF NERVE IMPULSES AND INITIATION OF MOTOR RESPONSES

In the central nervous system there are membranous elements that receive nerve impulses and initiate motor responses to these stimuli. At the biochemical level the nature of these membranous elements is still undefined and there is only guesswork as to the molecular basis of the operations in the command centers of the central nervous system.

The position is even more obscure in respect to the molecular basis of the processes by which sensory information is stored and recalled, and by which the organism can learn. But, the principle of the universality of the essential membrane modalities provides a rational basis for the hope that even these more complex events will in time be reducible to simple molecular processes.

THE SYSTEMS FOR CELL DIVISION

When a cell divides there is an elaborate system of membranous structures which controls the sequence of events by which the nuclear and cytoplasmic material is apportioned between the two daughter cells. Obviously, cell division involves major redistribution of parts, the breaking or disintegration of one set of membranes and the formation of new sets of membranes. There is a great deal of controlled and directed movement and this must call for specialized systems to execute this involved and vital function. It would be unprofitable to attempt to summarize the vast body of information already collected relating to the various parts and organelles of the mitotic apparatus. In terms of mechanistic principles, the mitotic apparatus poses no special problems. Some kind of contractile mechanism has to be invoked to account for the extensive directed movements which are intrinsic to cell division. The mechanisms by which existing membranes are disassembled and new membranes are assembled are not unique to the mitotic apparatus. We shall be describing other categories of membrane systems in which such extensive rearrangements of repeating units are observed. The special feature of the mitotic apparatus is to be found in the tactics by which the molecules of DNA are transformed from randomly dispersed forms to tightly coiled and compressed forms, and the tactics by which double-stranded DNA is separated into its component strands and by which each of the two strands is shepherded into the two daughter cells.

Cell division in amoeba does not involve the dissolution of the nuclear membrane. The segregation of chromosomes and the generation of the mitotic figures proceed within the nuclear membrane. It is, therefore, possible to recognize the nature of the mechanism by which the nuclear membrane aids and abets the process of cell division. Prior to cell division the nuclear membrane is spherical. At the actual point of cell division the nuclear membrane is shaped like a dumbbell, with the two sets of chromosomes and the asters apportioned between the two sectors of the dumbbell. This transition of the nuclear membrane corresponds to the shape transitions in the erythrocyte which are described in the following section. In the nonenergized state, the erythrocyte is spherical; in the energized state the erythrocyte has the shape of a biconcave disc which is geometrically analogous to a dumbbell. By analogy we may assume that during cell division in the amoeba the nuclear membrane becomes energized and undergoes transition to the dumbbell form—a form which facilitates the separation of the paired chromosomes and the asters. To complete the cycle the cell membrane undergoes furrow formation, and this furrow then bisects the nuclear membrane at the narrowest point of the dumbbell. The furrow formation of the cell membrane is probably also an expression of a nonenergized-to-energized transition triggered by the local production of ATP.

ENGULFING SYSTEMS

Multicellular and unicellular organisms have the capability for the membrane-mediated engulfing of particles external to a cell and the transport of the captured particle into the interior of the cell. Polymorphonuclear leukocytes engulf bacteria by a process known as phagocytosis. Protozoa, notably the amoeba, can also manipulate the plasma membrane to form tubular spurs, known as pseudopodia, which engulf particles external to the cell by phagocytosis. In cells

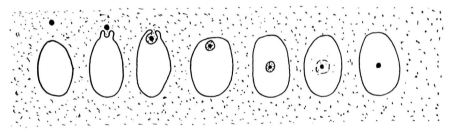

Fig. VIII.10. Diagrammatic representation of pinocytosis or phagocytosis.

generally, there is a process known as pinocytosis by which droplets of fluid, as well as solid particles, are transported into the interior. Within the cell, also there is the intracellular digestive system, the lysosomes—membranous organelles with a capability for engulfing intracellular material and for fusing with pinocytotic vacuoles. All these engulfing processes are basically variations on a single theme. A region of the plasma membrane is induced by an appropriate stimulus to evaginate a pouch which envelops a cell or particle or droplet. This envelopment continues to the point at which the ends of the rim of the hollowed pouch (see Figure VIII.10) meet and fuse. The pouch with the engulfed material, now inside the membrane, is then retracted into the interior of the cell and the pouch is pinched off

from the membrane to give rise to a pinocytotic vacuole. The important point is that a membrane surrounds the engulfed material until
it is disposed of. At that point the pinched off piece of vesicular
membrane can recoalesce with the plasma membrane.

Phagocytozing membranes need a fair amount of extra membranous
surface. Perhaps there are folded areas in the plasma membrane to
provide material for rapid evagination; there may be specialized
molecules either in the external medium or in the interior of the cell
which trigger the evaginating sequence. This sequence may be an
expression of a transition from a nonenergized to an energized
state—a transition that can permit a vesicular membrane to form
local membranous regions that are tubular.

The red blood corpuscle can exist in two forms—as a biconcave
disc and as a spherical unit. It has been conclusively demonstrated
that this shape change is controlled by the level of ATP. When ATP
is present, the red blood corpuscle assumes the shape of the biconcave
disc; when ATP is exhausted the red blood corpuscle rearranges to
form a spherical unit. The membrane of the red blood corpuscle,
even after removal of hemoglobin from the cells by washing under
appropriate conditions, shows comparable but not identical shape
changes in the transition from the energized state (in presence of
ATP) to the nonenergized state (removal of ATP). In the energized
state the membrane tends to bow in; in the nonenergized state the
membrane tends to bulge out. It is as though the basepieces were
concave out in the nonenergized state and convex in when energized.
The nonenergized membrane gives rise to exo vesicles which can be
pinched off, and these accumulate in the exterior. The energized
membrane gives rise to endo vesicles which can be pinched off, and
these accumulate in the interior. Thus, a large number of vesicles is
uniformly found *outside* the ghost in the nonenergized state, whereas
a correspondingly large number of vesicles is found *inside* the ghost
in the energized state of the membrane. In the erythrocytes of cattle,
the active transport of K^+ ions into the cell is a relatively slow process.
Thus, the ghosts of cattle erythrocytes are for all practical purposes
incapable of carrying on active transport of K^+. The complication
of the discharge of the energized state of the membrane by K^+ and/or
Na^+ is thus avoided in the study of shape changes in the ghosts of
cattle erythrocytes. What this means is that under appropriate

circumstances shape changes of the erythrocyte membranes are expressions, pure and simple, of the energizing and discharge of the plasma membrane. Thus, the phenomenon of pinocytosis may be looked upon as a property of all cell membranes. The essence of pinocytosis, pseudopodial action, and phagocytosis appears to be a cycle of local energizing and deenergizing of the membrane (extension and retraction). Some mechanism has to be involved for localized generation of ATP and then decay or removal of excess ATP after an appropriate interval.

SYSTEMS FOR TRANSPORT OF MACROMOLECULES

A wide variety of animal and plant cells are specialized for the task of secreting some special molecule be this a hormone, a mucoprotein, a neurotransmitter, or an enzyme. The glandular cells which secrete hormones, mucins, and digestive enzymes are all familiar examples of secretory cells but the complete list of secretory cells is surprisingly large. In all such secretory processes the Golgi apparatus plays a key role. The Golgi apparatus is a system of stacked membranes which are ultimately separable. At the bottom of the stack the membranes are flattened discs. At the top of the stack the membranes round up to form vesicles. Proteins synthesized by the endoplasmic reticulum are introduced into the discoid membranes at the bottom of the stack and this process continues until the membrane becomes engorged and is compelled to round up. The topmost membranes then fuse with the plasma membrane and leave the cell by pinocytosis. Outside the cell, these membranes perforate and the internal contents are released. There is a continuous cycle by which newly formed, discoid Golgi membranes are laid down at the bottom of the stack and engorged vesicular membranes leave the cell at the top of the stack. The membranes of the endoplasmic reticulum are lined up at the base of the stacked Golgi membranes. It has been shown by radioautography that the proteins which are secreted by the Golgi apparatus are synthesized in the endoplasmic reticulum and transferred by an as yet unknown mechanism to the discoid membrane at the bottom of the Golgi stack. Until very recently it was assumed that the Golgi apparatus is concerned exclusively with the task of transportation of specialized molecules from inside to outside the cell. However, studies of Neutra and Leblond have clearly established that the membranes of the Golgi apparatus contain sets of enzymes which carry out many of the enzymic reactions involved in the formation of the carbohydrate chains of glycoproteins. Moreover, it has become apparent that proteins secreted by the Golgi apparatus are invariably glycoproteins. The significance of this correlation has yet to be understood. Why should the proteins secreted by the Golgi apparatus have to be glycoproteins? What is the molecular logic inherent in the attachment of an extended carbohydrate residue to a protein? There is clearly an important molecular principle to be recognized.

PROTEIN SYNTHESIZING SYSTEMS

Many of the proteins of the cell are synthesized by ribosomes which are part of, or are associated with, the endoplasmic reticulum. The rationale of this membrane association is not fully understood (protein synthesis can certainly take place *in vitro* in the absence of membranes). Indeed, our present knowledge is rather incomplete with respect to the details by which complete proteins are synthesized molecule after molecule and then delivered to their ultimate location in the cell, and also with respect to the details by which sectors of membrane repeating units are assembled from their constituent proteins, and entire membranes are assembled from their repeating units. Clearly, there is some important implication in the fact that the nuclear synthesis of the informational molecules of the cell, as well as the ribosomal synthesis of the proteins which are determined by these informational molecules, almost invariably proceed within the precincts of membranes.

The traffic of messenger RNA from nucleus to endoplasmic reticulum is a feature of protein synthesis that still needs clarification. Furthermore, the movement of this messenger RNA along the polysome during protein assembly is a transducing process which has received very little attention. GTP is known to be hydrolyzed during protein assembly. One might speculate that the process is quite analogous to that already described for the actomyosin system. Perhaps interaction with GTP induces an energized conformational change resulting in a local extension of the RNA helix. Hydrolysis of the attached charged grouping might then allow the helix to reform, thus pulling the RNA molecule to a new position with respect to the particular ribosome at which the event occurred.

OTHER TRANSDUCING SYSTEMS

Although we have considered in some detail a number of important biological systems that transduce energy, the survey has by no means been exhaustive. Wherever there is energized movement there is *prima facie* evidence of molecular events involving ATP-energized conformational changes in collectivized transducing units.

SELECTED REFERENCES

BOOKS

Bourne, G. H., ed., "Structure and Function of Muscle," 3 Vols., Academic Press, New York, 1960: on the chemistry of myosin and actin.
Calvin, M., and Bassham, J. A., "The Photosynthesis of Carbon Compounds," Wm. Benjamin, Inc., New York, 1962.
Gest, H., San Pietro, A., and Vernon, L. P., eds., "Bacterial Photosynthesis," Antioch Press, Yellow Springs, 1964.
Green, D. E., and Goldberger, R. F., "Molecular Insights into the Living Process," Academic Press, New York, 1967.
Hodgkin, A. L., "The Conduction of the Nervous Impulse," Charles C Thomas, Springfield, Illinois, 1964: on nerve transmission.
J. Biophys. Biochem. Cytol., "The Sarcoplasmic Reticulum," Vol. 10, Supplement, Rockefeller University Press, New York, 1961.
Nachmansohn, D., "Chemical and Molecular Basis of Nerve Activity," Academic Press, New York, 1960: an introduction to the molecular mechanisms of the transmission of a nerve impulse.
"The Living Cell"; Readings from *Scientific American*, W. H. Freeman and Co., San Francisco, 1965: see article by D. Mazia on how cells divide; by W. H. Miller, F. Ratliff, and H. K. Hartline on how cells receive stimuli; by B. Katz on how cells communicate; and by A. Rich on polyribosomes.
Wolken, J. J., "Vision—Biophysics and Biochemistry of the Retinal Photo-receptors," Charles C Thomas, Springfield, 1966: an invaluable guide to the intricacies of the visual system.

SPECIAL ARTICLES

Arnon, D. I., *Physiol. Rev.* **47** (3), 317 (1967): on photosynthesis.
Arnon, D. I., Tsujimoto, H. Y., and McSwain, B. D., *Nature* **214**, 562 (1967): on ferredoxin.

Beams, H. W., and Kessel, R. G., *Intern. Rev. Cytol.* **23**, 209 (1968): on the structure and function of the Golgi apparatus.

Boardman, N. K., and Anderson, J. M., *Biochim. Biophys. Acta* **143**, 187 (1967): on the isolation of particles from spinach chloroplasts with different photosynthetic capabilities.

Branton, D., and Park, R. B., *J. Ultrastruct. Res.* **19**, 283 (1967): on the ultrastructure of the chloroplast.

Bridges, C. D. B., *in* "Comprehensive Biochemistry" (M. Florkin and E. H. Stotz, eds.), Vol. 27, p. 31, Elsevier, New York, 1967: on the biochemistry of visual processes.

Clayton, R. K., *in* "Bacterial Photosynthesis,", p. 377, Antioch Press, Yellow Springs, Ohio, 1963: on the photochemical reaction centers in photosynthetic tissues.

Dallner, G., *Acta Pathol. Microbiol. Scand.*, Suppl. 166, 1 (1963): on the endoplasmic reticulum.

Davies, R. E., *in* "Essays in Biochemistry" (P. N. Campbell and G. D. Greville, eds.), Vol. 1, p. 29, Academic Press, New York, 1965: on the mechanism of muscular contraction.

De Robertis, E., *J. Gen. Physiol.* **43**, Suppl. 2, 1 (1960): on the ultrastructure of the retinal rod system.

Ebashi, S., and Endo, M., *Progr. Biophys. Mol. Biol.* **18**, 123 (1968): calcium ion and muscular contraction.

Fuller, R. C., and Nugent, N. A., *Proc. Natl. Acad. Sci., U.S.* in press (1969): on pteridines and their role in the photosynthetic reaction center.

Hanson, J., and Lowy, J., *J. Mol. Biol.* **6**, 46 (1963): on the chemistry of myosin and actin.

Hasselbach, W., and Makinose, M., *Biochem. Biophys. Res. Commun.* **7**, 132 (1962): translocation of Ca^{2+} by the sarcoplasmic reticulum.

Heller, J., *Biochemistry* **7**, 2906; **7**, 2914 (1968): on the isolation of the visual pigment and on the conformational changes on light exposure in the visual pigment.

Holter, H., *Intern. Rev. Cytol.* **8**, 481 (1959).

Howell, S. H., and Moudrianakis, E. N., *Proc. Natl. Acad. Sci., U.S.* **58**, 1261 (1967): on the identification of the ATPase of chloroplast with headpieces.

Huxley, H. E., *Sci. Am.* **213**, 18 (1965): on the mechanism of muscular contraction.

Huxley, H. E., *Harvey Lectures Ser.* **60**, 85 (1966): on the mechanism of muscular contraction.

Huxley, H. E., *in* "The Cell" (J. Brachet and A. E. Mirsky, eds.), Vol. IV, Academic Press, New York, 1968: on the mechanism of muscular contraction.

Huzisige, H., Usiyama, H., Kikuti, T., and Azi, T., *Plant Cell Physiol. (Tokyo)* **10**, 441 (1969): on the isolation of two photoactive particles from spinach chloroplasts.

Jamieson, J. D., and Palade, G. E., *J. Cell Biol.* **34**, 577 (1967): on the Golgi apparatus.

Karlin, A., and Bartels, E., *Biochim. Biophys. Acta* **126**, 525 (1966): on the relation of sulfhydryl groups to the acetylcholine-activated permeability system of the electroplax.

Kirk, J. T., and Tilney-Bassett, R. A. E., *in* "The Plastids: Their Chemistry Structure, Growth, and Inheritance," Chapter 1, Freeman, London, 1967: on the ultrastructure of the chloroplast.

Neutra, M., and Leblond, C. B., *J. Cell Biol.* **30**, 119 (1966): on the synthesis of the carbohydrate of mucus in the Golgi apparatus of goblet cells.

Neutra, M., and Leblond, C. B., *Sci. American* **220**, 100 (1969): on the Golgi apparatus.

Olson, J. M., Filmer, D., Radloff, R., Romano, C. A., and Sybesma, C., *in* "Bacterial Photosynthesis," p. 423, Antioch Press, Yellow Springs, Ohio, 1963: on the protein-chlorophyll-770 complex from green bacteria.

Palade, G. E., *in* "Electron Microscopy in Anatomy" (J. D. Boyd, F. R. Johnson, and J. D. Lever, eds.), p. 176; Edward Arnold, London, 1961: on the endoplasmic reticulum.

Palade, G. E., and Siekevitz, P., *J. Cell Biol.* **34**, 577 (1962); *J. Mol. Biol.* **19**, 507 (1966): on the Golgi apparatus.

Park, R. B., and Pheifhofer, A. O. A., *Proc. Natl. Acad. Sci., U.S.* **60**, 337 (1968): on the localization of the quantasome.

Parker, C. J., and Gergely, J., *J. Biol. Chem.* **236**, 411 (1961): on the relaxing factor.

Penniston, J. T., and Green, D. E., *Arch. Biochem. Biophys.* **128**, 339 (1968): on the energized configurational changes in the membrane of the red blood corpuscle during pinocytosis.

Perry, S. V., *Progr. Biophys. Mol. Biol.* **17**, 325 (1967): on the chemistry of actomyosin.

Remsen, C. C., Valois, F. W., and Watson, S. W., *J. Bacteriol,* **94**, 422 (1967): on the internal membrane systems of *Nitrocystis oceanus.*

San Pietro, A., *Ann. Rev. Plant Physiol.* **16**, 155 (1965): on photosynthesis.

Skou, J. C., *Physiol. Rev.* **45**, 596 (1965): on Na^+- and K^+-activated ATPase.

Stein, W. D., *Biochem. J.* **105**, 3P (1967): on permeases.

Von Hippel, P. H., Gellert, M. F., and Morales, M. F., *Conf. Chem. Muscular Contraction, Tokyo, 1957*, Igaku Shoin, Ltd., Tokyo, 1958, p. 1: on actomyosin.

Wald, G., *Ann. Rev. Biochem.* **22**, 497 (1953): on the biochemistry of visual processes.

KEY REFERENCES FOR CHAPTER VIII

Universal principles of energy transduction

Green and Goldberger (1967), Chapter 15

Actomyosin and the contractile system

Davies (1965), Ebashi and Endo (1968), Hanson and Lowy (1963),

Huxley (1965, 1966, 1968), Perry (1967), Von Hippel *et al.* (1958)

Chloroplast system

Arnon (1967), Arnon *et al.* (1967), Branton and Park (1967), Boardman and Anderson (1967), Clayton (1963), Fuller and Nugent (1969), Howell and Moudrianakis (1967), Huzisige *et al.* (1969), Kirk and Tilney-Bassett (1967), Olson *et al.* (1963), Park and Pheifhofer (1968), San Pietro (1965)

Membranes carrying out active transport

Hasselbach and Makinose (1962), Parker and Gergely (1961), Skou (1965), Stein (1967)

Permeases

Stein (1967)

Membranes concerned with sensory reception

Bridges (1967), De Robertis (1960), Heller (1968), Wald (1953)

Membranes concerned with nervous excitation

Hodgkin (1964), Karlin and Bartels (1966), Nachmansohn (1960)

Cell division

Mazia (1965)

Pinocytosis

Holter (1959), Penniston and Green (1968)

Golgi apparatus

Neutra and Leblond (1969), Palade and Siekevitz (1962)

Endoplasmic reticulum and protein synthesis

Dallner (1963), Palade (1961), Rich (1965)

APPENDIX I

BIOENERGETICS

Certain fundamental concepts have been taken for granted in this book. The logical development of the description of the mitochondrion would otherwise have been distorted by lengthy parenthetical detours. Thus, we have assumed that the reader is familiar with the metabolic pathways of intermediary metabolism and with the chemistry of such key substances as heme proteins and adenine nucleotides. We have also assumed a familiarity with those aspects of chemistry and thermodynamics which relate to bioenergetics. Since, however, the principles of bioenergetics underlie the whole subject matter of this book, we shall develop in this Appendix a short elementary treatment of some of these principles for the benefit of those readers not already well versed in them.

WATER

When water is formed by a combination of hydrogen and oxygen there is a release of a considerable amount of energy. Each of the O—H bonds formed in this reaction has a partially polar character; the H end of the bond bears a fractional positive charge, and the oxygen atom bears the corresponding fractional negative charge. The overall result is a water molecule with two characteristic dipoles at a bond angle with respect to each other which is approximately tetrahedral. The wave mechanical considerations which account for the precise magnitude and direction of these dipoles are not relevant to our

159

discussion, but the realities of the dipole and the resulting capacity for hydrogen bonding are apparent in the high dielectric constant of water, its unique characteristics as a solvent, and its high boiling point, latent heat, and surface tension. These properties, in turn, account for the role of water as the milieu for most living processes. The extended, hydrogen bonded, lattice structure of water, present to some degree even in its liquid state, also underlies hydrophobic bonding between nonpolar molecules in an aqueous environment. Nonpolar chains immobilize large volumes of the surrounding water into "icebergs" (the only structures which allow for maximal hydrogen bonding *around* nonpolar groups) and thus tend to increase the free energy of the system (by decreasing the entropy). Dissolution of these icebergs (resulting in states having the lowest free energy and the greatest stability) is achieved by nonpolar groupings clumping together. Thus, the dipole of water accounts, indirectly, for the interaction between repeating units of membranes. At lower temperatures, iceberg structures imposed by isolated nonpolar chains do not involve so large a decrease in entropy as they do at higher temperatures; hence, the clumping of nonpolar groupings as a tactic for the dissolution of the icebergs does not involve such a large fall in free energy. In other words, hydrophobic bonds become weaker at lower temperatures, a fact which accounts for the phenomenon of "cold inactivation" of enzymes composed of multiple subunits held together by hydrophobic bonding.

OXIDATIONS AND REDUCTIONS

The dipole of the O—H bond is important in another respect. It illustrates the "electronegativity" of the oxygen atom—in other words, the high affinity of the oxygen atom for electrons. This affinity is also expressed in the energy which is released upon the formation of water when the two unpaired electrons in the oxygen atom become paired with electrons from two hydrogen atoms. The dipole of the O—H bond also illustrates another important facet, i.e. the relatively *low* "electronegativity" of the hydrogen atom (that is to say, the relative ease with which this atom can donate its electron to an appropriate acceptor).

The addition of oxygen (or any similarly electronegative atom or grouping) to any other atom will tend to draw electrons from that atom, hence, it is reasonable that either the addition of oxygen *or* the removal of electrons should be considered to be formally equivalent processes. Both are classified as *"oxidations."* It also follows that, because hydrogen atoms tend to donate electrons to the atoms with which they are bonded, the removal of hydrogen atoms with their full complement of electrons effectively denudes these atoms of electrons, hence "oxidizes" them. Indeed, the removal of two hydrogen atoms may be looked upon as an ionization with the release of a free proton followed by the removal of a hydride anion (a proton with *two* electrons in a filled 1-s orbital). Conversely, it can be seen that the removal of an oxygen atom (with only its own complement of electrons) or the addition of two hydrogen atoms are both equivalent to the addition of electrons. All three operations can therefore be classified together under the same heading as *"reductions."*

PHOTOSYNTHESIS AND RESPIRATION

The carbon atom is only slightly more electronegative than is hydrogen; the C—H bond is therefore usually covalent, with almost equal sharing of the bonding electrons. Now, the essence of photosynthesis is the reduction of CO_2 by water; i.e., the removal of electrons (and their associated protons) from their very stable association (in water) with a very electronegative atom (oxygen) and their transfer to carbon, which, as we have just seen, does not have a high affinity for electrons. Photosynthesis is, therefore, endergonic. Respiration involves the oxidation of organic foodstuffs which derive from reduced CO_2, i.e. molecules which contain C—H and C—C bonds. The essence of respiration, therefore, is the exergonic transfer of electrons (together with associated protons) from the relatively electropositive carbon atom back to the electronegative oxygen atom.

OXIDATION-REDUCTION COUPLES

So far, the energetic concepts involved in biological oxidations have been considered in a purely qualitative manner. Indeed, an understanding of the functioning of the mitochondrion requires no more than a general qualitative understanding of the *nature* of the chemical energy involved. It is, however, useful to consider in a more nearly quantitative manner the magnitude of this form of energy.

Oxidations can usually be considered in terms of a pair of reversible reactions or couples. Thus, the reaction $Cu + 0.5O_2 \rightarrow CuO$ may be broken down into the two half reactions:

$$Cu \rightleftharpoons Cu^{2+} + 2e^- \tag{1}$$

$$O + 2e^- \rightleftharpoons O^{2-} \tag{2}$$

Similarly, the reaction $Cu^{2+} + Zn \rightarrow Cu + Zn^{2+}$ is the resultant of the couples:

$$Zn \rightleftharpoons Zn^{2+} + 2e^- \tag{3}$$

$$Cu^{2+} + 2e^- \rightleftharpoons Cu \tag{4}$$

It will be seen that in the first example copper is the reductant (and is itself oxidized), whereas in the second reaction the cupric ion is the oxidant (and is itself reduced to copper). Thus, the direction in which the half-reaction $Cu \rightleftharpoons Cu^{2+} + 2e^-$ proceeds depends upon the other half-reaction to which it is coupled. If this second half-reaction involves a stronger oxidizing agent than the cupric ion, then the reaction $Cu \rightleftharpoons Cu^{2+} + 2e^-$ will be driven to the right. Conversely, the reaction will be driven to the left if the other half-reaction involves a poorer oxidizing agent than the cupric ion.

Now, for a pair of half-reactions of this kind to proceed, the reactants do not necessarily have to be mixed; each reaction can take place in a separate "half-cell" provided that electrical contact is made between these half-cells. For example, the reaction

$$[2Cu^{2+} + H_2 \rightarrow 2Cu^+ + 2H^+]$$

is the sum of the reactions:

$$2Cu^{2+} + 2e^- \rightleftharpoons 2Cu^+ \tag{5}$$

$$H_2 \rightleftharpoons 2H^+ + 2e^- \tag{6}$$

If a mixture of cupric and cuprous ions is placed in contact with an inert metal electrode (e.g. platinum), reaction (5) can proceed in either direction as electrons are either supplied or removed at the electrode. Similar considerations would apply to a H_2/H^+ half-cell, i.e. one in which hydrogen gas is bubbled over a platinum electrode immersed in an acidic solution. If electrical contact is made between these two half-cells, then, because hydrogen is a better reducing agent than is the cuprous ion, reaction (6) will proceed to the right. That is to say, electrons will be removed at the H_2/H^+ electrode, flow along the wire connecting the two half-cells, and thus be supplied at the Cu^+/Cu^{2+} electrode to drive reaction (5) to the right.

If the "hydrogen electrode" in this example were replaced by a half-cell containing a mixture of ferric and ferrous ions, then the electrons would flow in the opposite direction. The ferric ion is a better oxidizing agent than is the cupric ion; in other words, the ferric:ferrous couple has a greater tendency to remove electrons from the electrode than has the cupric:cuprous half-cell. The overall reaction which proceeds, therefore, is:

$$Cu^+ + Fe^{3+} \rightarrow Cu^{2+} + Fe^{2+} \qquad \qquad \textbf{(7)}$$

FREE ENERGY OF OXIDATION-REDUCTION REACTIONS

It can now be seen that the direction of a particular oxidation-reduction reaction will be manifested by the direction of electron flow measurable between the two half-cells, i.e. by the sign of the potential difference across the pair of electrodes as measured by a simple voltmeter. Now, if a known amount of charge (Q) is allowed to move between two electrodes with a potential difference of E, the work performed is simply the quantity ($Q \times E$).* If *one equivalent* of reactant is oxidized, F coulombs of negative charge must be moved from one half-cell to the other (F is the Faraday constant, 9.65×10^4 coulomb mole^{-1}). Under these circumstances the work performed is equivalent to the free energy change in the reaction. Hence, this free energy change can be calculated from the potential difference between the electrodes by the simple relationship:

$$\varDelta G \text{ (change in Gibbs free energy/mole)} = -F \times E \qquad \textbf{(8)}$$

In general, if n electrons are involved in the particular reaction considered, $\varDelta G = -nFE$. If E is expressed in volts, then, since 1 volt-Faraday $= 96,500$ joules and 4.18 joules $= 1$ cal:

$$\varDelta G = -nE \times 23,000 \text{ cal/mole} \qquad \textbf{(9)}$$

Hence, if the potential difference (E) across a pair of half-cells is known, the free energy of the oxidation-reduction reaction can

* If Q is measured in coulombs and E in volts, the work units are joules.

readily be calculated in terms of calories per mole. This free energy change is related to the equilibrium constant for a reaction. Knowledge of E under standard conditions, therefore, should allow us to predict the point at which a given oxidation-reduction reaction will come to equilibrium.

ELECTRODE POTENTIALS

It is apparent that any half-cell reaction couple (such as $H_2:2H^+$; $Cu^+:Cu^{2+}$) will have a characteristic tendency ("electrode potential") to donate electrons to an electrode or to remove them from an electrode. The actual direction of electron flow, the actual free energy change, and the final position of equilibrium in any overall oxidation-reduction reaction will depend upon the differences between the "electrode potentials" of the couples involved in the particular reaction. Actual measurements of potentials can give us only the potential *difference* between a pair of half-cells; so, in order, to assign an absolute number to each half-cell potential, we must place an arbitrary value on the potential of one half-cell (preselected to serve as a suitable standard) and measure the potentials of all other half-cells with reference to the potential of the standard. The standard chosen is a half-cell in which hydrogen gas at a pressure of one atmosphere is in equilibrium with a solution of hydrogen ions of unit activity (pH $= 0$). A platinum electrode in such a half-cell is said to have a standard electrode potential, E_0, of zero. Any couple with a greater tendency than this standard half-cell to donate electrons to an electrode (hence, leave it negatively charged) is said to have a more negative standard electrode potential. (This would apply to strong reducing agents; the more electro*positive* a species, the more *negative* will be the electrode potential it will set up.) Conversely, any system with a greater tendency than the $H_2:H^+$ couple to withdraw electrons from the electrode (hence, to leave it positively charged) will have a positive standard electrode potential. (Thus, strongly electro*negative*, oxidizing agents have very *positive* electrode potentials.)

Once the standard electrode potential of a half-cell, E_0^A, has been measured against the standard hydrogen electrode ($E_0 = 0$), it can itself be used as the standard for measuring the standard potential (E_0^B), of another half-cell, since the measured potential difference (E) between the cells will be the difference between E_0^A and E_0^B. If E_0^A and E_0^B are both known, then E can be calculated; hence, the free energy change of the oxidation-reduction reaction between the two systems, were they allowed to interact, can also be calculated from the simple equation derived above:

$$\Delta G = -nE \times 23,000 \text{ cal/mole} \tag{10}$$

STANDARD OXIDATION-REDUCTION POTENTIALS

Standard electrode potentials are defined with all reactants at unit activity. Now it is clear from elementary considerations of the principles of mass action that a relative decrease in the concentration of the oxidized species of a given oxidation-reduction couple would lead to an increased tendency toward production of the oxidized species, i.e. to a more negative electrode potential. For example, if the H^+ concentration at a hydrogen electrode were decreased 100 times, so that hydrogen gas at one atmosphere were bubbled through a solution at pH 2 (instead of at pH 0), there would be a greatly increased tendency for the reaction $H_2 \rightarrow 2H^+ + 2e^-$ to be driven to the right, and the electrode would become more negative. In general, the observed potential, E_{obs}, is given by the equation:

$$E_{obs} = E_0 + \frac{RT}{nF} \ln \frac{[\text{oxidized form}]}{[\text{reduced form}]} \tag{11}$$

where n is the number of electrons involved in the reaction, and the brackets denote the molar concentrations of the two forms. (In the example given, the reduced form is in excess, and so the logarithmic correction factor becomes negative in sign.) This equation, however, does not cover those cases in which the reduced form is a molecule containing hydrogen atoms which are released as hydrogen ions upon oxidation, for example, the hydroquinone-quinone couple:

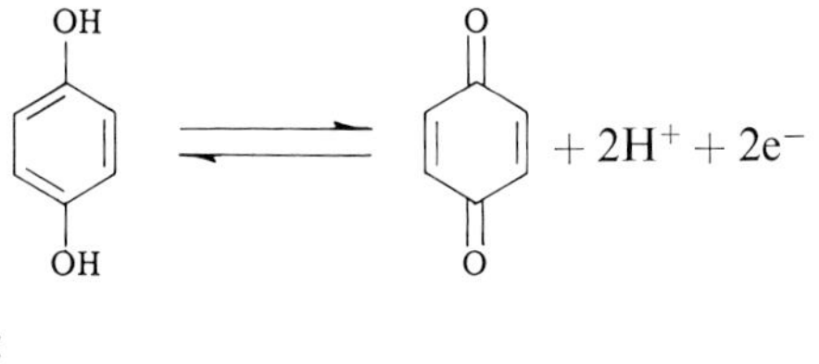

$$\tag{12}$$

In such cases:

$$E_{obs} = E_0 + \frac{RT}{nF} \ln \frac{[\text{oxidized form}]}{[\text{reduced form}]} + x \frac{RT}{nF} \ln [H^+] \tag{13}$$

where E_0 is the potential when the oxidized and reduced forms, and (H^+), are all at unit activity and x is the number of protons released per molecule oxidized. Thus, whenever protons are released upon oxidation, the observed electrode potential is dependent upon pH. Now, many of the components involved in mitochondrial electron

TABLE A1
Standard Oxidation-Reduction Potentials

System	E'_0 (pH 7) volts*
Oxygen/water	0.82
Cytochrome a_3; oxidized/reduced	0.53
Cytochrome a ; oxidized/reduced	0.29
Cytochrome c; oxidized/reduced	0.26
Cytochrome c_1; oxidized/reduced	0.22
Cytochrome b; oxidized/reduced	0.00
Coenzyme Q; oxidized/reduced	0.00
Fumarate/succinate	0.03
f_D; oxidized/reduced (flavoprotein of Complex I)	−0.22
f_S; oxidized/reduced (flavoprotein of Complex II)	−0.22
Lipoic acid; oxidized/reduced	−0.29
$DPN^+/DPNH$	−0.32
$TPN^+/TPNH$	−0.32
H^+/H_2	−0.42
Ferredoxin; oxidized/reduced	−0.43
Succinate/α-ketoglutarate	−0.67
Acetate/pyruvate	−0.70

*Certain of the values quoted are approximations based upon conflicting published data, frequently obtained under different conditions.

transfer do release protons on oxidation; so, even under conditions in which there is an equal concentration of oxidized and reduced forms, the relevant oxidation-reduction potential will be far removed from the corresponding E_0, because these reactions are carried out at physiological pH and not at pH 0. For this reason, it has proved useful to define a "standard oxidation-reduction potential," E_0', corresponding to the electrode potential under conditions in which the oxidized and reduced forms are at equal activity and the pH is

7.0. From tables of E_0' for biological oxidation-reduction couples (see Table A.1) the free energy change for reactions at pH 7.0 between any pair of such couples can immediately be calculated. [It will be noted that in such tables the hydrogen electrode acquires a very negative potential, since [H^+] is now 10^{-7} instead of 1.0.]

PROTON RELEASE AND ELECTRON TRANSFER

The newcomer to this field is frequently confused by the apparent transition at Complex III of the electron transfer chain between a "hydrogen transfer chain" and a simple electron transfer chain. Such difficulties arise because the couples involved in respiration are of three kinds:

(i) $X^{2+} \rightleftharpoons X^{3+} + e^-$ (e.g. cytochrome c)   (14)

(ii) $YH + H^+ \rightleftharpoons Y^+ + 2H^+ + 2e^-$ (e.g. DPN)   (15)

(iii) $ZH_2 \rightleftharpoons Z + 2H^+ + 2e^-$ (e.g. coenzyme Q or fumarate)   (16)

Now these formulations are rather arbitrary. Whether, for example, couples of type (iii) should actually be written as:

$$ZH_2 \rightleftharpoons Z + (2H), \text{ or}$$ (17)

$$ZH_2 \rightleftharpoons Z + H^- + H^+$$ (18)

or whether couples of type (ii) should be written as:

$$YH \rightleftharpoons Y^+ + H^-$$ (19)

are really questions concerned with the actual mechanism of the oxidative reaction. Thus, the overall reaction might be of the kind:

$$[ZH_2 + Y^+ \rightleftharpoons Z + YH + H^+]$$ (20)

which could be considered as the sum of equations (15) and (16) or, equally well, as the sum of equations (18) and (19). On the other hand, the overall reaction might involve no net proton release, as in:

$$[ZH_2 + A \rightleftharpoons Z + AH_2],$$ (21)

or the release of two protons as in:

$$[ZH_2 + 2X^{3+} \rightleftharpoons Z + 2X^{2+} + 2H^+]$$ (22)

For simplicity, however, it may be considered that all dehydrogenations are formally equivalent to the removal of a pair of electrons plus a pair of protons. The electron acceptor may then be considered, upon reduction, to take up both, one, or neither of these protons, depending upon the nature of the acceptor molecule.

FREE ENERGY OF BIOLOGICAL OXIDATIONS

E_0' at pH 7.0 for the dehydrogenation of many organic metabolic intermediates (ZH_2/Z) is of the order of -300 mV. E_0' for $H_2O/\frac{1}{2}O_2$ at the same pH is $+820$ mV. The difference in potential between these couples is thus greater than 1.1 volt, and since 2 electrons are involved in the overall reaction, $ZH_2 + \frac{1}{2}O_2 \rightleftharpoons Z + H_2O$, the fall in free energy, $-\Delta G'$, is approximately $1.1 \times 2 \times 23{,}000$ cal/mole, i.e. about 50,000 cal/mole.* In general, the transfer of a pair of electrons between two couples whose oxidation-reduction potentials differ by 220 mV is equivalent to a standard free energy change of about 10,000 cal/mole.

It can be seen from Table A.1 that each of the reactions catalyzed by Complexes I, III, and IV of the mitochondrial electron transfer chain (DPNH–Q, QH_2–ferricytochrome c, and ferrocytochrome c–O_2 reductase, respectively) involves a change in oxidation-reduction potential of more than 220 mV. It can also be seen that the α-ketoglutarate-succinate couple has a substantially more negative potential than does the DPNH–DPN$^+$ couple. Are these differences sufficient to account for the synthesis of ATP during either oxidative phosphorylation or substrate level phosphorylation? How much energy is stored in a "high energy bond," and what is the nature of this energy?

"HIGH ENERGY BONDS"

There is no real mystery as to what constitutes a "high energy bond." When the bond X—Y is broken, say by hydrolysis, to yield the products XH and YOH, the energy released will depend upon the difference between the free energy of the products and that of the reactants. When this difference in free energy is unusually high, as in the hydrolysis of ATP to ADP and inorganic phosphate, the

---

* $\Delta G'$ is the standard free energy change when all the reactants are at unit concentration at pH 7.0. This entity is frequently given the designation $\Delta F'$.

bond which is broken is conventionally referred to as being of "high energy." Thus, the "energy" of a bond in the sense that we are using it here is the free energy change accompanying hydrolysis of the bond. This free energy change, which gives us a measure of the amount of work the reaction could perform under isothermal conditions, is related to the equilibrium constant of the reaction:

$$\varDelta G^0 = -RT \ln K. \tag{23}$$

Now $\varDelta G^0$ in this equation is the standard free energy change, when 1 mole of reactant is converted into 1 mole of product under conditions in which each is maintained at unit activity. Since ATP, ADP, and phosphate are all ionizing species, the actual free energy change, $\varDelta G$, will vary with pH. At pH 7.0 the reaction is approximately represented as:

$$ATP^{4-} + H_2O \rightarrow ADP^{3-} + HPO_4^{2-} + H^+ \tag{24}$$

Under these conditions (at 25°) the free energy change, when ATP, ADP, and phosphate are all at concentration 1.0 M, is considered to be about -7000 cal/mole. In the cell, however, many factors affect the actual free energy of hydrolysis of ATP. The key factors are the actual concentrations of the reactants and products, the nature of the intramitochondrial buffers,* the fact that ATP, ADP, and phosphate all form complexes of differing stability with intracellular Mg^{2+}, and the fact that the pH in isolated compartments within the cell (if indeed "pH" has a useful meaning in a very small compartment), may be significantly greater or less than 7.0.

In view of all these considerations, it should be borne in mind that the quoted $\varDelta G$ of -7000 cal/mole for the hydrolysis of ATP probably does not represent the actual free energy of the "high energy bond" under the conditions existing during mitochondrial oxidative phosphorylation. A value of -9000 cal/mole has been taken to be a more realistic "working" figure. Thus, under conditions appropriate to the working of the mitochondrion, and with coupling of high efficiency, a pair of electrons when falling in oxidation-reduction potential by about 200 mV would provide sufficient energy to synthesize a high energy bond.

---

* In view of the proton produced in reaction (24), the heat of buffer ionization can make a very significant contribution to the overall $\varDelta G$.

NATURE OF "HIGH ENERGY BOND"

There are three factors which contribute to the unusually high free energy of hydrolysis of bonds such as the pyrophosphate bonds of ATP. One is the actual electronic configuration of the bond itself, which, as has been shown by recent wave mechanical analysis, results in the bridging oxygen atoms bearing a fractional *positive* charge. A second factor is the unusually low free energy of the products of hydrolysis. The nonionized form of the terminal pyrophosphate bond of ATP can be written as:

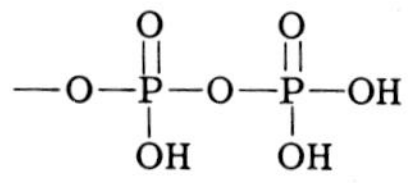

$$-O-\overset{\displaystyle O}{\overset{\displaystyle \|}{\underset{\displaystyle OH}{P}}}-O-\overset{\displaystyle O}{\overset{\displaystyle \|}{\underset{\displaystyle OH}{P}}}-OH$$

Upon hydrolysis the nonionized forms of the products would be:

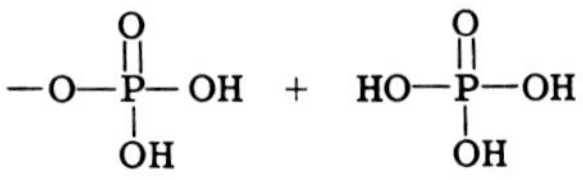

$$-O-\overset{\displaystyle O}{\overset{\displaystyle \|}{\underset{\displaystyle OH}{P}}}-OH \quad + \quad HO-\overset{\displaystyle O}{\overset{\displaystyle \|}{\underset{\displaystyle OH}{P}}}-OH$$

Thus, two new —OH groups are produced, each of which increases the number of possible resonating forms of the ionized product molecules. Resonance-stabilized ionization therefore takes place. (At physiological pH this hydrolysis results in the production of one full equivalent of hydrogen ion.) As a result of this stabilization by resonance, the hydrolysis products are at a level of free energy significantly lower than is the parent molecule. From this point of view the high energy pyrophosphate bonds of ATP are strictly analogous to any other acid anhydride bonds.

A third factor which increases the free energy of hydrolysis is the negative charge carried by each of the two products (ADP^{3-} and HPO_4^{2-}); these charges cause repulsion between the product molecules, hence tend to drive the equilibrium further in favor of hydrolysis. To the extent that this repulsive factor contributes to $\Delta G'$, the latter might be affected if the hydrolysis were to take place in an environment whose dielectric constant differed greatly from that of water.

SELECTED REFERENCES

BOOKS

Clark, W. M., "Oxidation-Reduction Potentials of Organic Systems," Williams and Wilkins, Baltimore, Maryland, 1960.

Edsall, J. T., and Wyman, J., "Biophysical Chemistry," Academic Press, New York, 1958.

Klotz, I. M., "Energy Changes in Biochemical Reactions," Academic Press, New York, 1967: an inspired introduction to biological energetics.

Lehninger, A. L., "Bioenergetics," W. A. Benjamin, New York, 1965.

SPECIAL ARTICLES

Huennekens, F. M., and Whiteley, H. R., in "Comparative Biochemistry" (M. Florkin and H. S. Mason, eds.), Vol. I, pp. 107–186, Academic Press, New York, 1960: on energy-rich compounds.

Mahler, H. R., and Cordes, E. H., in "Biological Chemistry," Chapter V, Harper and Row, New York, 1966.

APPENDIX II

FORMULAS

Acetyl-CoA[*]

$$CH_3COSCoA$$

*where CoASH represents coenzyme A (see below).

Acetylcholine

$$\underset{\text{CH}_3\text{C}}{\overset{\text{O}}{\parallel}}-O-CH_2CH_2\overset{+}{N}(CH_3)_3$$

(*cis*) Aconitate: the anion of

$$CH_2COOH$$
$$|$$
$$C \cdot COOH$$
$$\parallel$$
$$CH \cdot COOH$$

The double bond allows for *cis-trans* isomerism. The form arising from the citric acid cycle is the anion of *cis*-aconitic acid.

ADP (adenosine 5'-diphosphate)

See ATP.

L-Alanine

$$\underset{\text{CH}_3\text{CH} \cdot \text{COOH}}{\overset{\text{NH}_2}{|}}$$

AMP (adenosine 5'-monophosphate)

See ATP.

173

Antimycin A

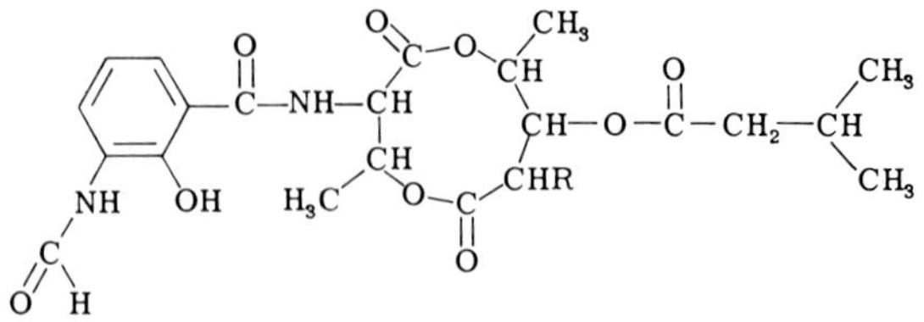

$R = C_6H_{13}$ in antimycin A_2

$ = C_4H_9$ in antimycin A_3

L-Ascorbic acid (L-ascorbate = anion)

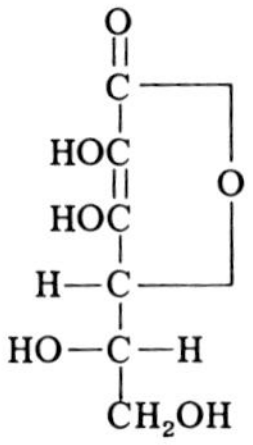

α-L-Aspartic acid

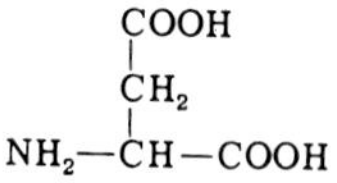

ATP (adenosine 5′-triphosphate)

The anion of:

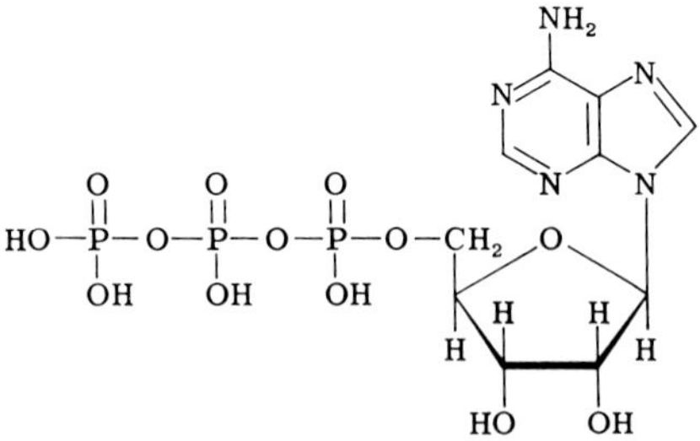

9-β-D-Ribofuranosyl adenine 5′-triphosphate

AMP and ADP have mono- and diphosphate groups on the 5′ position, respectively.

Atractylate (atractyloside)

The potassium salt of:

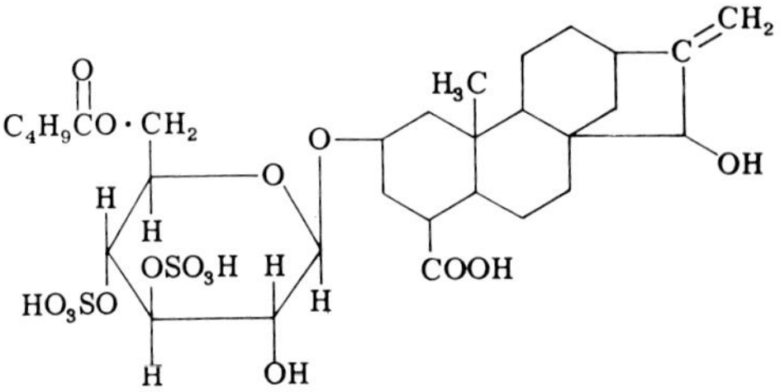

Azide

$$^-N{=}\overset{+}{N}{=}N^-, \ N_3^-$$

Carbonyl cyanide phenylhydrazones

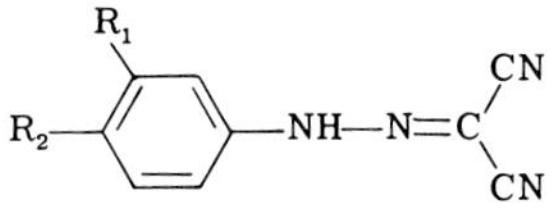

When R_1 = H and R_2 = F_3CO: carbonyl cyanide p-trifluoromethoxyphenylhydrazone.

When R_1 = Cl and R_2 = H: carbonyl cyanide m-chlorophenylhydrazone (m-ClCCP).

Cardiolipin

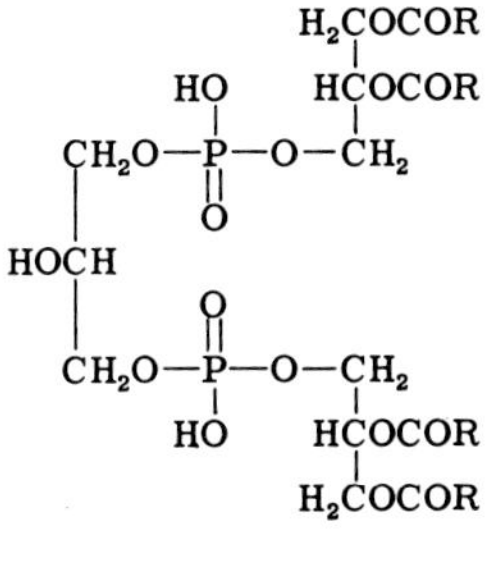

where RCO = acyl group.

Carnitine (3-hydroxy-4-trimethylaminobutyric acid)

$$(CH_3)_3\overset{+}{N}{-}CH_2^-{-}CH{-}CH_2^-{-}COO^-$$
$$\underset{OH}{|}$$

Chlorophylls *a* and *b*

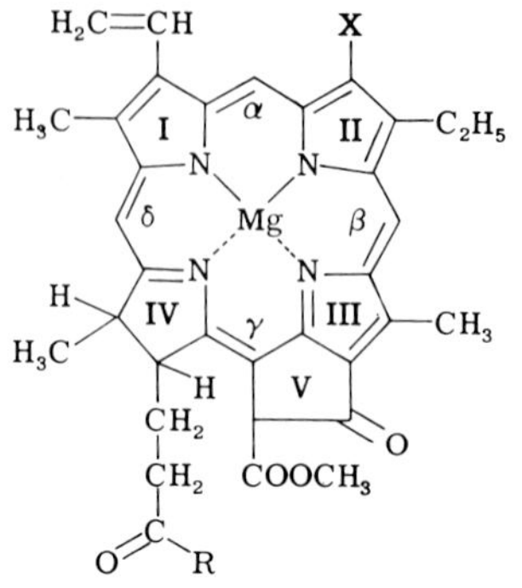

When X = —CH₃ the structure is chlorophyll *a* .

When X = —C⟨O/H the structure is chlorophyll *b* .

R = phytyl:

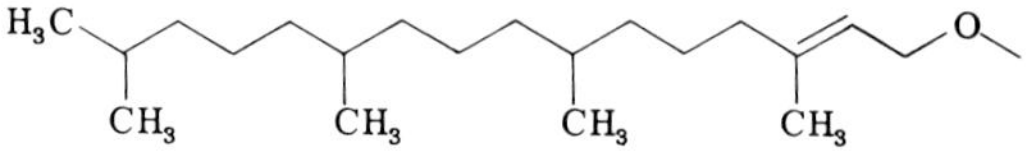

Cholesterol

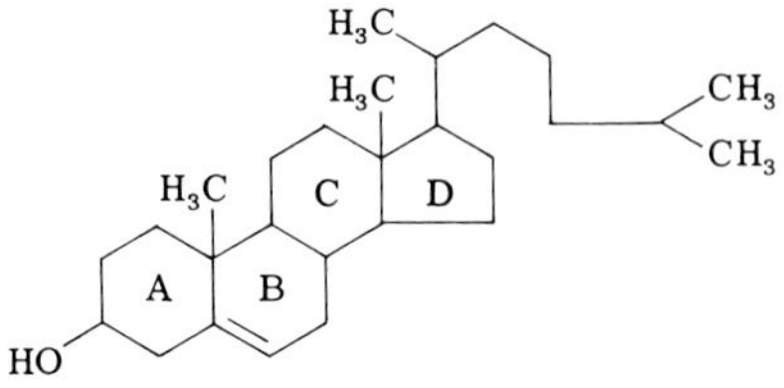

Citric acid (citrate = anion)

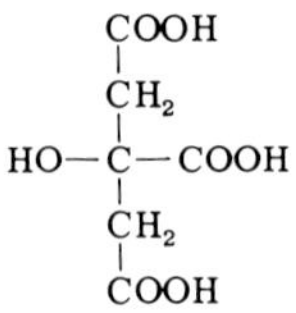

CoASH (coenzyme A)

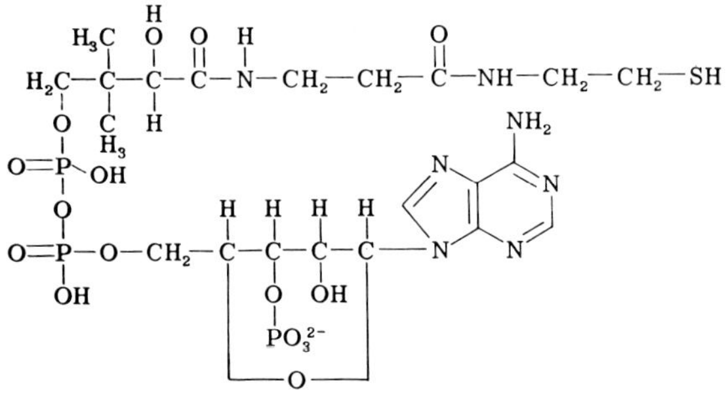

CoQ (coenzyme Q, ubiquinone)

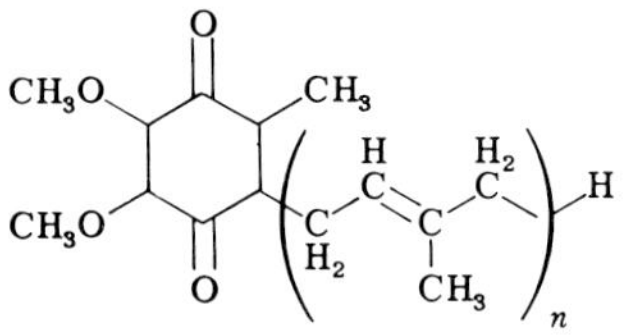

n = 10 in CoQ from beef heart, (CoQ_{10})

CTP (cytidine 5'-triphosphate)

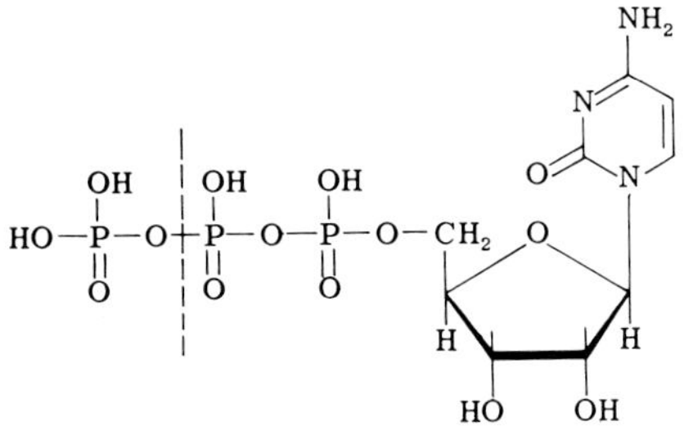

Cytidine diphosphatocholine (CDP-choline)

See CTP; the terminal phosphate of CTP is replaced by choline:

$(CH_3)_3\overset{+}{N}CH_2{-}CH_2O{-}$

Cytidine diphosphatoethanolamine (CDP-ethanolamine)

See CTP; the terminal phosphate of CTP is replaced by ethanolamine:

$NH_2{-}CH_2{-}CH_2{-}O{-}$

Cytochromes

For cytochromes other than cytochrome *c* (below) see Hemoproteins.

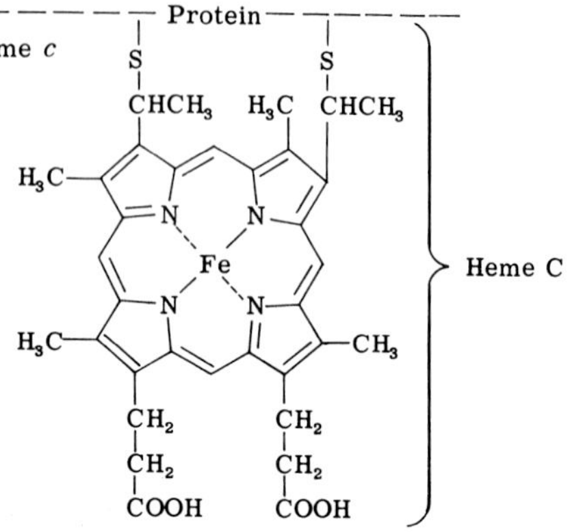

Amino acid sequence of human heart cytochrome *c*:

Acetyl-*N*-Gly-Asp-Val-Glu-Lys-Gly-Lys-Lys-Ile-Phe-Ile-Met-Lys-Cys-Ser-Glu(NH₂)-Cys

└──heme──┘

His-Thr-Val-Glu-Lys-Gly-Gly-Lys-His-Lys-Thr-Gly-Pro-Asp(NH₂)-Leu-His-Gly-Leu-
Phe-Gly-Arg-Lys-Thr-Gly-Glu(NH₂)-Ala-Pro-Gly-Tyr-Ser-Tyr-Thr-Ala-Ala-Asp(NH₂)-
Lys-Asp(NH₂)-Lys-Gly-Ile-Ile-Trp-Gly-Glu-Asp-Thr-Leu-Met-Glu-Tyr-Leu-Glu-
Asp(NH₂)-Pro-Lys-Lys-Tyr-Ile-Pro-Gly-Thr-Lys-Met-Ile-Phe-Val-Gly-Ile-Lys-Lys-
Lys-Glu-Glu-Arg-Ala-Asp-Leu-Ile-Ala-Tyr-Leu-Lys-Lys-Ala-Thr-Asp(NH₂)-GluCOOH

Dicumarol (dicoumarol; 3, 3'-methylenebishydroxycoumarin)

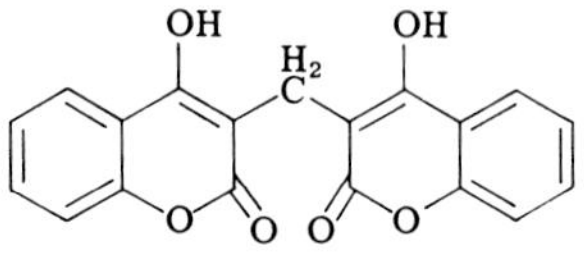

2, 4-Dinitrophenol

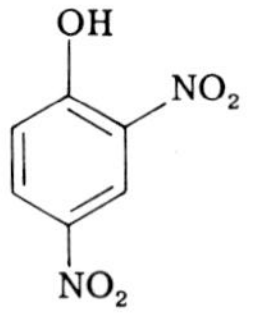

Dithionite

$$S_2O_4^{2-}$$

DNA (deoxyribonucleic acid)

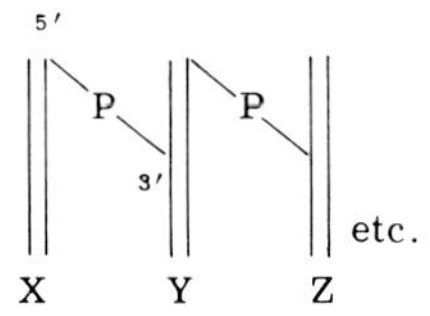

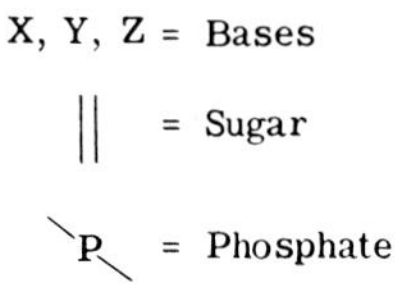

X, Y, Z = Bases

$\|$ = Sugar

$\diagdown$P$\diagdown$ = Phosphate

The bases are:

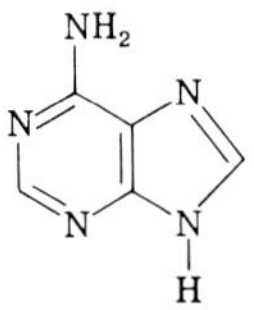

Adenine

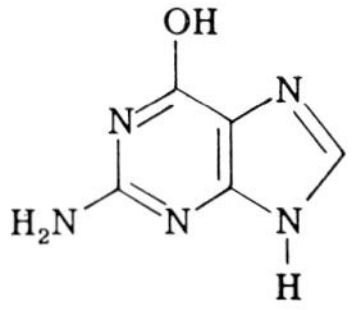

Guanine

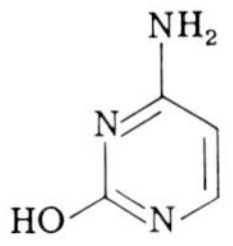

Cytosine

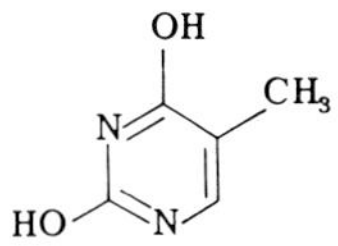

Thymine

The sugar is 2-deoxy D-ribose, linked to the base by an N-glycosidic bond:

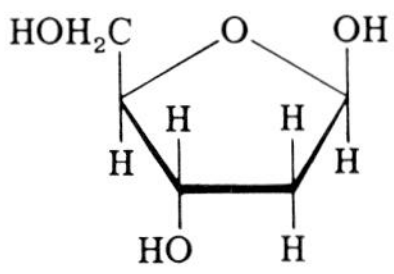

The phosphate group is derived from orthophosphoric acid:

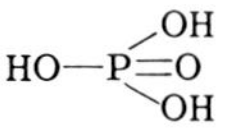

$$HO-P\substack{\diagup OH \\ =O \\ \diagdown OH}$$

RNA (ribonucleic acid)

This has similar structure to DNA, except that the bases involved do not usually include thymine, but uracil instead.

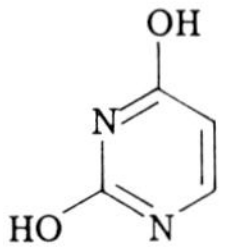

$$
\begin{array}{c}
\text{OH} \\
\text{HO} \quad \text{N} \quad \text{N}
\end{array}
$$

The sugar involved is D-ribose:

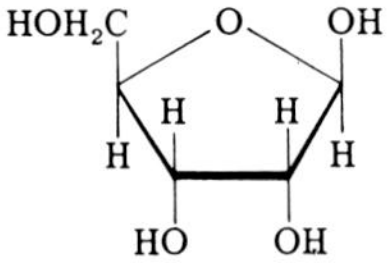

$$
\text{HOH}_2\text{C} \quad \text{O} \quad \text{OH}
$$

DPN$^+$ (diphosphopyridine nucleotide, nicotinamide adenine dinucleotide, NAD$^+$;
(coenzyme I)

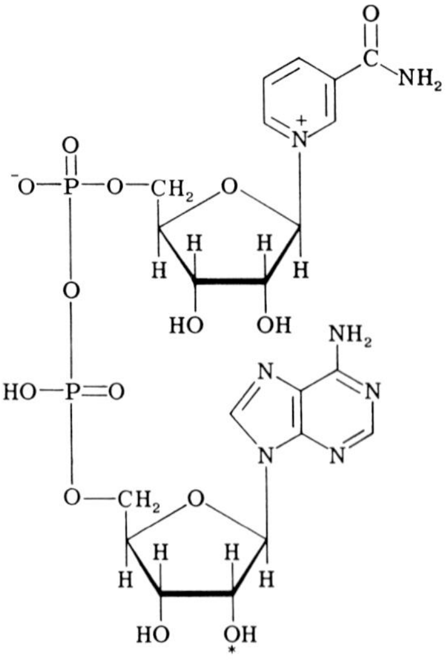

$$
\text{O}=\text{P}\text{—O—CH}_2 \qquad \text{O}
$$

$$
\text{HO—P}=\text{O}
$$

$$
\text{O—CH}_2 \qquad \text{O}
$$

*Another phosphate group is added here for TPN$^+$ (NADP$^+$: coenzyme II).

DPNH (NADH)

As for the corresponding oxidized pyridine nucleotide except for the pyridine ring itself:

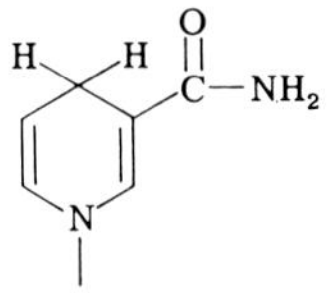

(similarly for TPNH, NADPH)

FAD (flavin adenine dinucleotide)

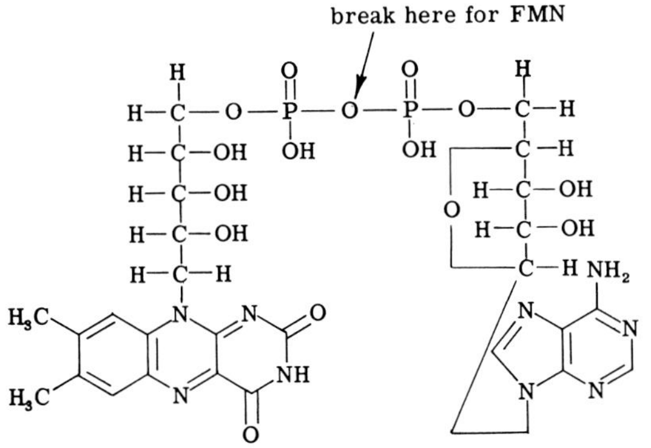

break here for FMN

In the reduced form of FAD, the isoalloxazine ring takes up 2 hydrogen atoms:

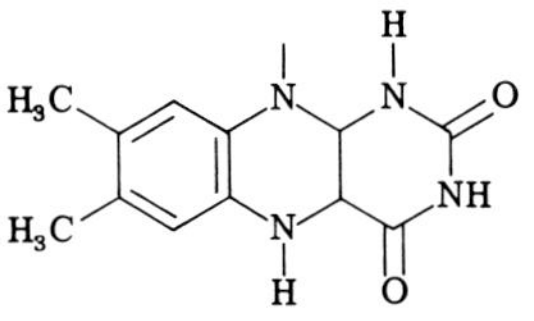

Ferredoxin (Fd)

A nonheme-iron protein:

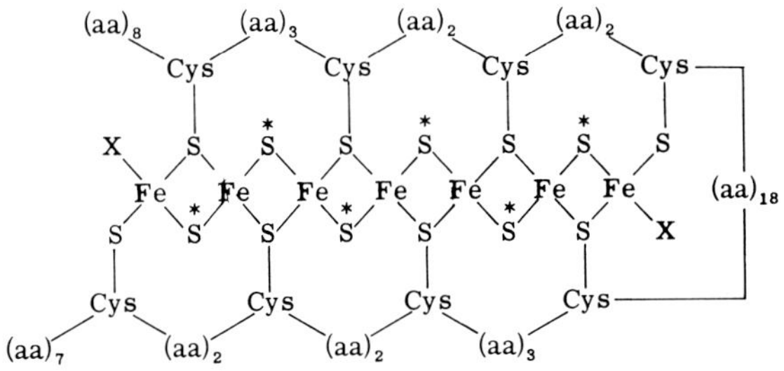

aa = Amino acyl residues   S* = "Acid labile S," i.e. sulfide

Flavoprotein

  Specific proteins attached to a flavin prosthetic group (FAD or FMN)

FMN (flavin mononucleotide, riboflavin)
  See FAD.

Fumarate

  The anion of fumaric acid:

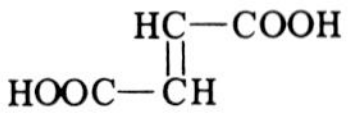

$$
\begin{array}{c}
\text{HC}-\text{COOH} \\
\parallel \\
\text{HOOC}-\text{CH}
\end{array}
$$

α-D-Glucose

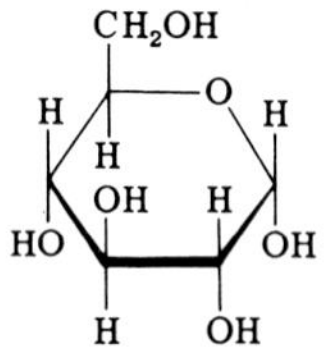

Glutamic acid (glutamate = anion)

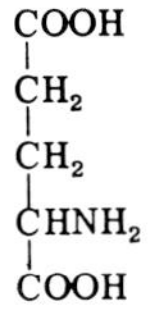

$$
\begin{array}{l}
\text{COOH} \\
| \\
\text{CH}_2 \\
| \\
\text{CH}_2 \\
| \\
\text{CHNH}_2 \\
| \\
\text{COOH}
\end{array}
$$

Glycerol phosphoryl-glycerol

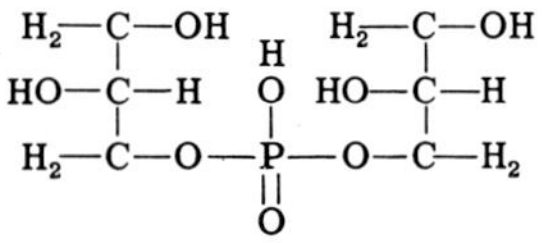

α-Glycerophosphate: the anion of

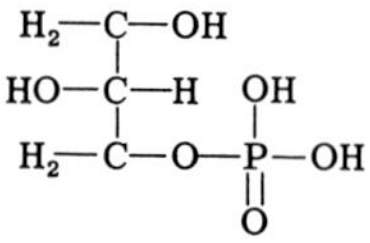

GDP (guanosine 5′-diphosphate)

See ADP; the structure is the same except that the base is guanine:

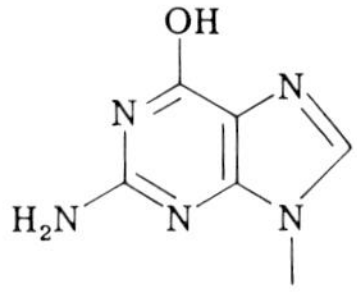

OH

H_2N

GMP (guanosine 5′-monophosphate)

See GDP; same structure but with only one phosphate group.

Gramicidin S

Isolated from a tryothrycin mixture.

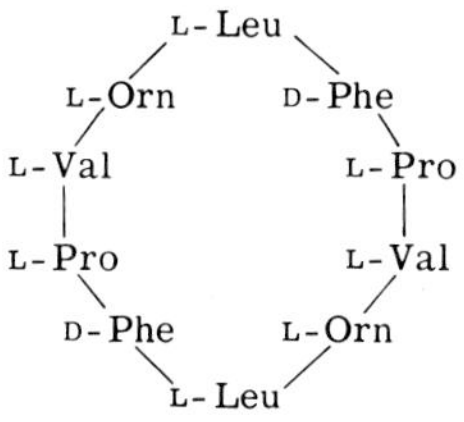

L-Leu
L-Orn   D-Phe
L-Val   L-Pro
L-Pro   L-Val
D-Phe   L-Orn
L-Leu

Gramicidin A contains: D-Leu; L-Trp; DL-Val; L-Ala; Gly.

GTP

See ATP; same structure, but with guanine as base instead of adenine. See also GDP.

Guanidine

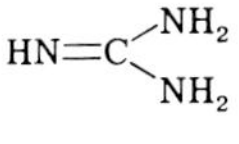

$HN{=}C\langle\begin{smallmatrix}NH_2\\NH_2\end{smallmatrix}$

Hemes (haems)

Square planar chelates of Fe in which the metal can form complexes with two additional ligands.

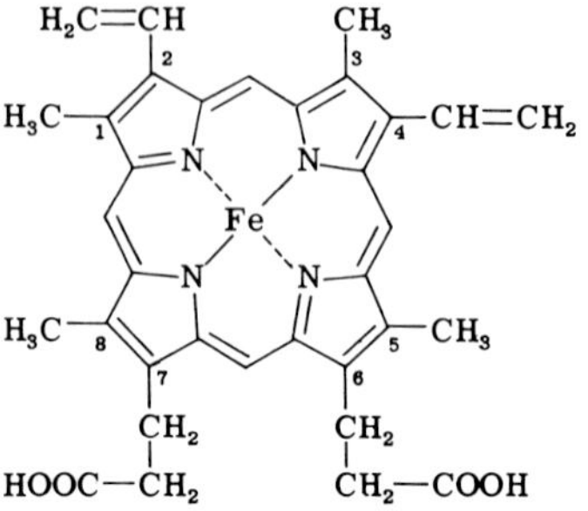

Ferroprotoporphyrin IX
(proto) heme (IX)

When the substituent at 2 is $R-S-\overset{\displaystyle CH_3}{\underset{\displaystyle |}{C}}H$, and that at 4 is $-\overset{\displaystyle CH_3}{\underset{\displaystyle S-R}{C}}H$ (where R—SH is a

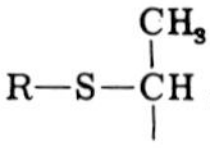

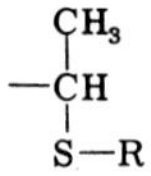

cysteinyl residue of a hemoprotein), then the heme is heme C.

When the substituent at 2 is $HO-\overset{\displaystyle |}{\underset{\displaystyle |}{C}}H$ $(CH_2)_3-\overset{|}{\underset{CH_3}{C}}H-(CH_2)_3-\overset{|}{\underset{CH_3}{C}}H-(CH_2)_3-\overset{CH_3}{\underset{CH_3}{C}}H$ and that at

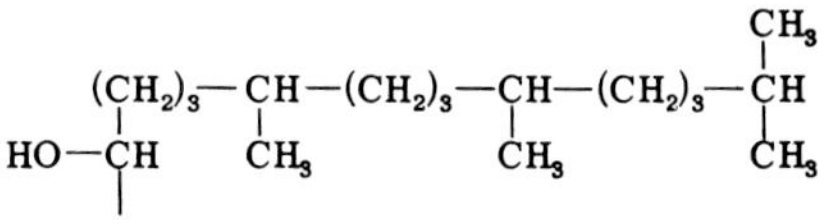

8 is —CHO   then the heme is heme A.   (this latter structure is tentative.)

Hemoproteins

Proteins whose prosthetic group is a heme, attached by coordination at one or both of the positions on the Fe at right angles to the plane of the heme molecule. (Attachment in the case of proteins containing heme C is also covalent through cysteine residues; see cytochrome c above).

Hemoproteins include:
hemoglobin (protoheme IX); myoglobin (protoheme IX); catalase (protoheme IX); cytochrome a (heme A); cytochrome a_3 (heme A); cytochrome b (protoheme IX); cytochrome b_5 (protoheme IX); cytochrome b_6 (protoheme IX); cytochrome c (heme C); cytochrome c_1 (heme C); cytochrome f (heme C); and cytochrome h (heme C).

β-Hydroxybutyric acid

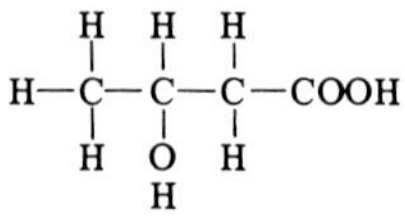

$$H-\overset{\displaystyle H}{\underset{\displaystyle H}{C}}-\overset{\displaystyle H}{\underset{\displaystyle O}{C}}-\overset{\displaystyle H}{\underset{\displaystyle H}{C}}-COOH$$

Both the L, (+) and the D, (–) forms are of metabolic significance.

Isocitrate

    The anion of isocitric acid:

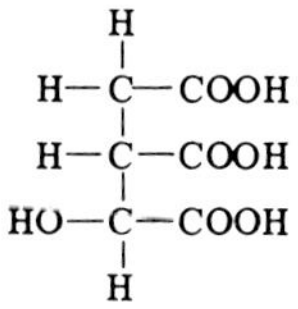

$$
\begin{array}{c}
\text{H} \\
| \\
\text{H}-\text{C}-\text{COOH} \\
| \\
\text{H}-\text{C}-\text{COOH} \\
| \\
\text{HO}-\text{C}-\text{COOH} \\
| \\
\text{H}
\end{array}
$$

α-Ketoglutarate (α-oxoglutarate)

    The anion of:

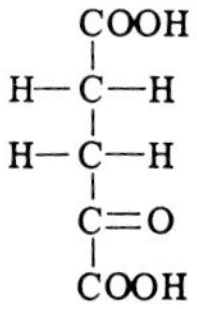

$$
\begin{array}{c}
\text{COOH} \\
| \\
\text{H}-\text{C}-\text{H} \\
| \\
\text{H}-\text{C}-\text{H} \\
| \\
\text{C}=\text{O} \\
| \\
\text{COOH}
\end{array}
$$

β-Lactose (4-D-glucopyranosyl-β-D-galactopyranoside)

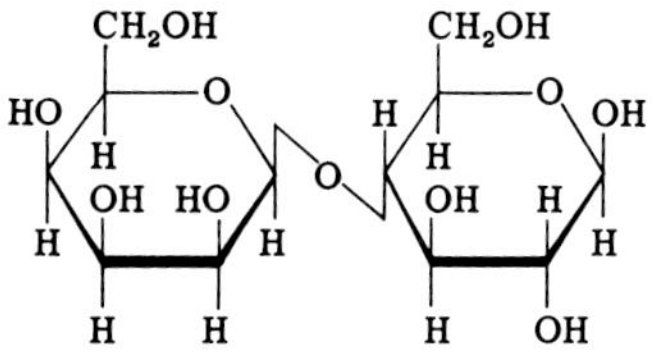

α-Lecithin (phosphatidylcholine)

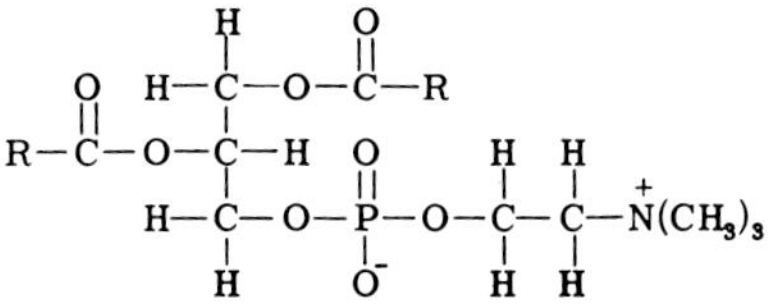

where RCO = acyl group.

    β-Lecithin has choline phosphate on the β position of the glycerol moiety.

Lipoic acid (thioctic acid; 6, 8-dithio-octanoic acid)

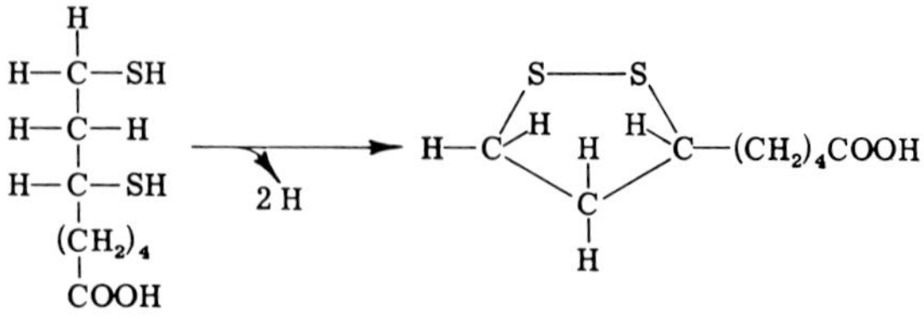

(reduced)   (oxidized)

L-α-Lysine

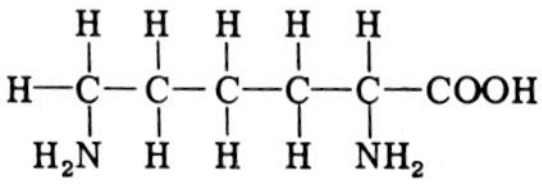

Malate

   The anion of malic acid:

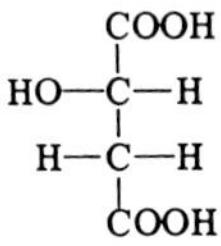

L-α-Methionine

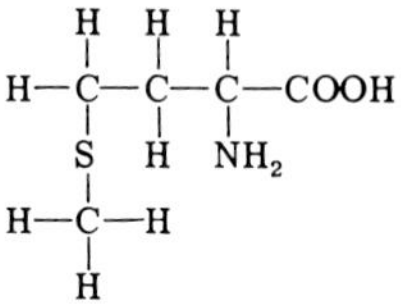

Myristate

   The anion of:

$$CH_3(CH_2)_{12}COOH$$

Nigericin

$$C_{39}H_{69}O_{11}$$

Nonheme iron

In the context of the respiratory electron transfer mechanism, "nonheme iron" refers to the several-fold excess of iron atoms over the porphyrin-containing proteins present in the mitochondrion. A portion of the iron atoms not chelated by porphyrins is associated with each of the submitochondrial complexes I, II, and III, and participates in a structure which on reduction gives rise to an EPR signal at g = 1.94. This signal disappears on reoxidation of the complexes. Whether and where this nonheme iron structure participates in the electron transfer-oxidative phosphorylation mechanism is a matter of current investigation. The nonheme iron structure giving rise to the g = 1.94 signal occurs widely in redox enzyme systems, e.g. the mitochondria, the nitrogen-fixing apparatus (in bacterial ferredoxins), chloroplasts (in plant ferredoxin), the steroid and related mixed function oxidases (in adrenodoxin, testodoxin, and putidaredoxin), and the metalloflavoproteins such as xanthine oxidase, aldehyde oxidase, and dihydroorotic dehydrogenase. In each case, the nonheme iron structure appears to accept and pass on only one electron at a time. From studies on the plant and mixed function oxidase types, it appears that the smallest functional unit of "nonheme iron" (yielding the g = 1.94 signal on reduction) may be a pair of vicinal iron atoms complexed to two sulfur atoms (which appear as sulfide in acid media) and presumably to other ligands donated by the protein and/or the solvent.

L-Noradrenaline (norepinephrine)

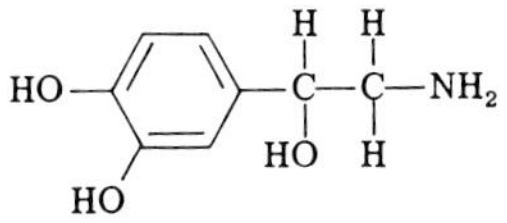

Oligomycin

Antibiotic complex produced by an actinomycete.

A: $C_{24}H_{40}O_6$

B: $C_{22}H_{36}O_6$

C: $C_{28}H_{46}O_6$

Orthophosphoric acid (the various anions of which are designated as P_i)

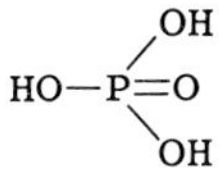

Ouabain (G-strophanthin)

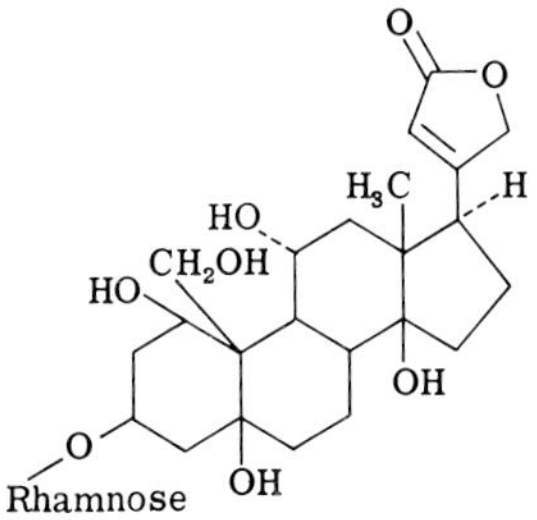

Oxaloacetate (oxosuccinic acid, ketosuccinic acid)
    The anion of:

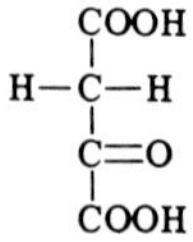

$$\begin{array}{c} COOH \\ | \\ H-C-H \\ | \\ C{=}O \\ | \\ COOH \end{array}$$

Palmitate
    The anion of:

$$CH_3(CH_2)_{14}COOH$$

Performic acid

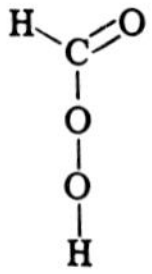

Phosphatidylcholine
    See lecithin.

Phosphatidylethanolamine (α-cephalin)

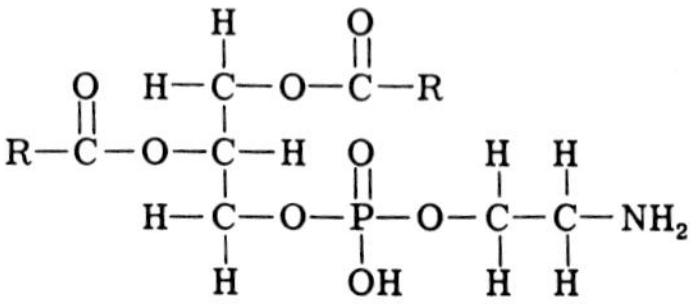

RCO = Acyl group

Phosphatidylglycerol

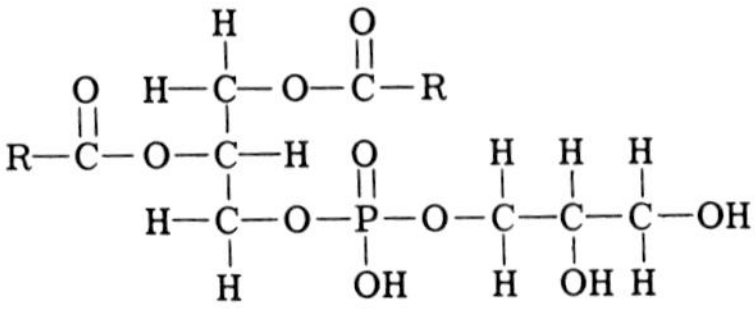

1-Phosphatidyl-L-*myo*-inositol

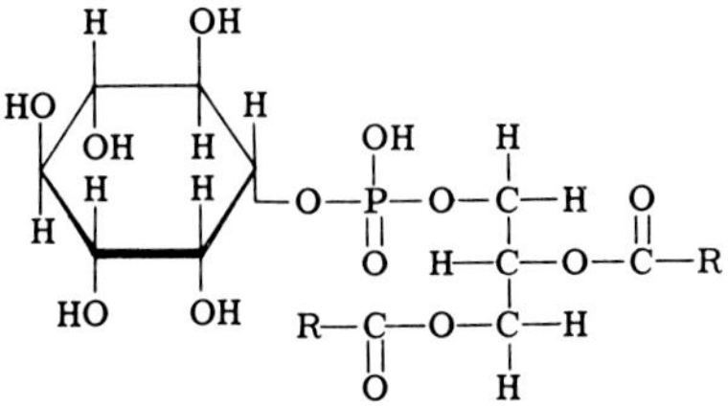

RCO = Acyl group

    myo-Inositol often has phosphate groups on OH's.

1-Phosphatidylserine

    Same as 1'-phosphatidyl-L-*myo*-inositol, except that there is a serine residue instead of *myo*-inositol.

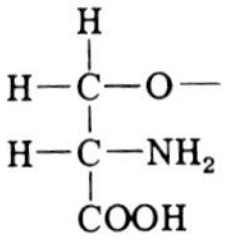

Phosphoryl

    The ionized forms of the radical:

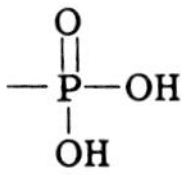

Plastoquinone

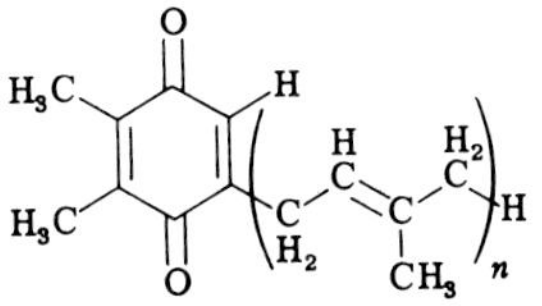

Polylysine

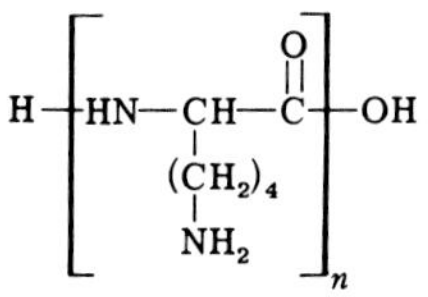

Polyunsaturated fatty acids

For example: Arachidonic acid

$$CH_3(CH_2)_4-(CH=CHCH_2)_4-(CH_2)_2-COOH$$

or   Linolenic acid

$$CH_3CH_2CH=CHCH_2CH=CHCH_2CH=CH(CH_2)_7-COOH$$

L-α-Proline

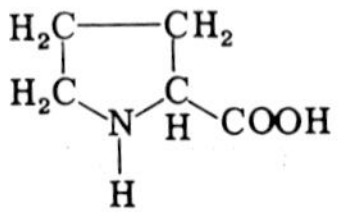

Propionate

The anion of:

$$CH_3CH_2COOH$$

Pyrophosphate (PP$_i$)

The anion of:

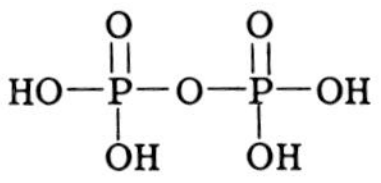

Pyruvate

The anion of:

$$CH_3COCOOH$$

Retinal (vitamin A aldehyde, retinene)

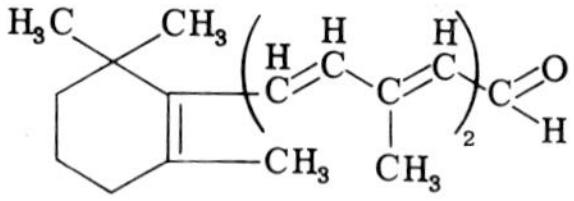

When aldehyde group is replaced by
   —CH$_2$OH = retinol, vitamin A alcohol

   —COOH = retinoic acid

Of possible geometric isomers, the all-*trans*- and 11-*cis* configurations are the ones of physiological interest.

Rhodopsin (visual purple)

   Visual pigment formed by combination of retinal and the protein opsin.

RNA (ribonucleic acid)

   See DNA.

Rotenone

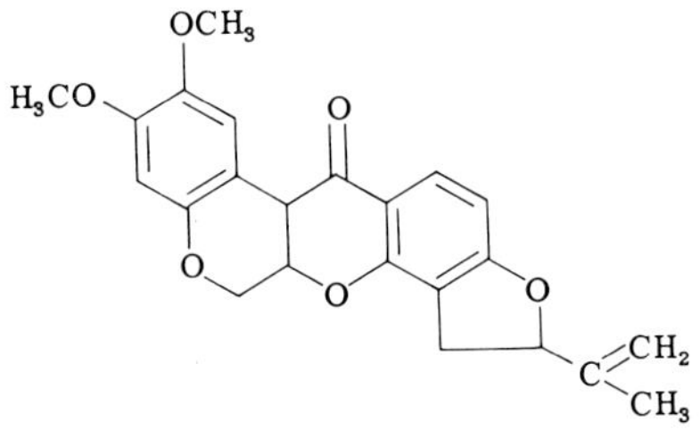

Rutamycin

   See oligomycin.

Sphingomyelin

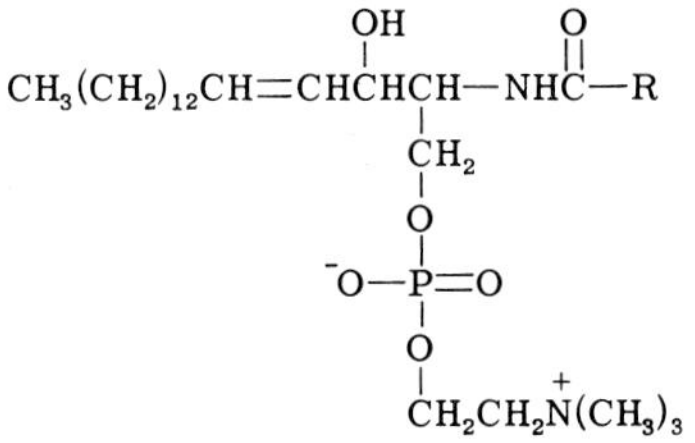

   Natural form is *trans*, *erythro*, i.e., —OH and —NH—groups on opposite sides.

   RCO = Acyl

Stearate

   Anion of:

$$C_{17}H_{35}COOH$$

Succinate

  Anion of :

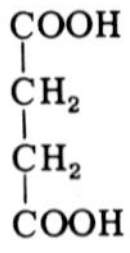

$$
\begin{array}{c}
\text{COOH} \\
| \\
\text{CH}_2 \\
| \\
\text{CH}_2 \\
| \\
\text{COOH}
\end{array}
$$

Succinyl-CoA

$$
\begin{array}{c}
\text{COOH} \\
| \\
\text{CH}_2 \\
| \\
\text{CH}_2 \\
| \\
\text{COSCoA}
\end{array}
$$

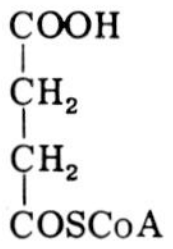

For the structure of CoASH, see previous entry CoASH.

Sucrose (1-α-D-glucopyranosyl-β-D-fructofuranoside)

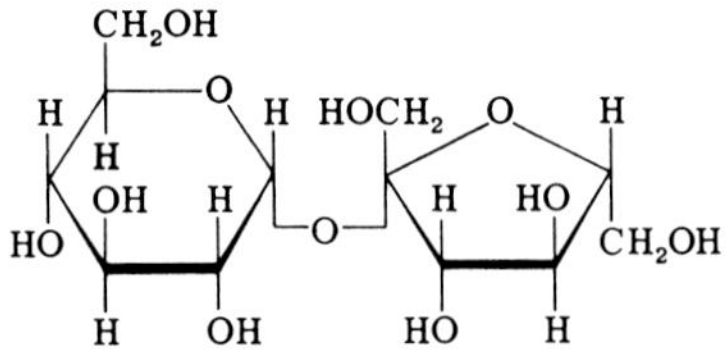

Taurocholate

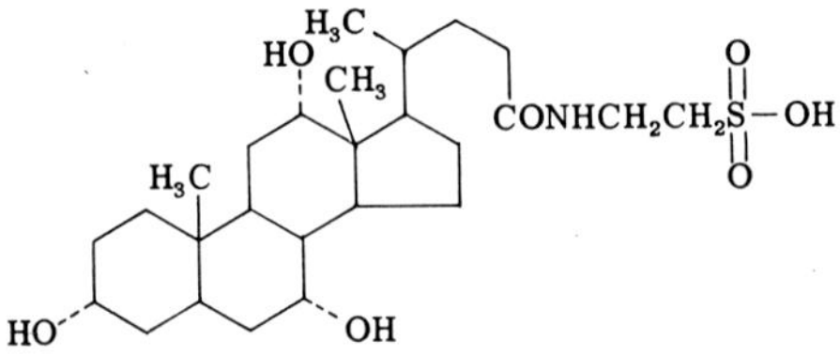

TPP (thiamine pyrophosphate)

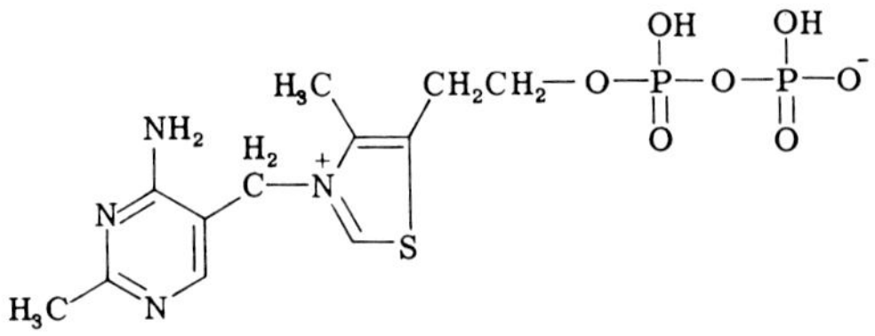

TMPD (N,N,N',N'-tetramethyl-p-phenylenediamine)

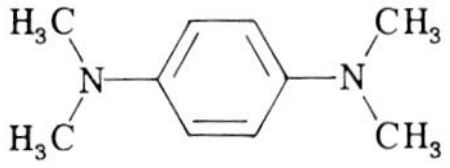

TPN$^+$, TPNH

    See DPN$^+$, DPNH.

Urea

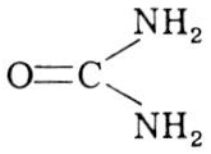

Valinomycin

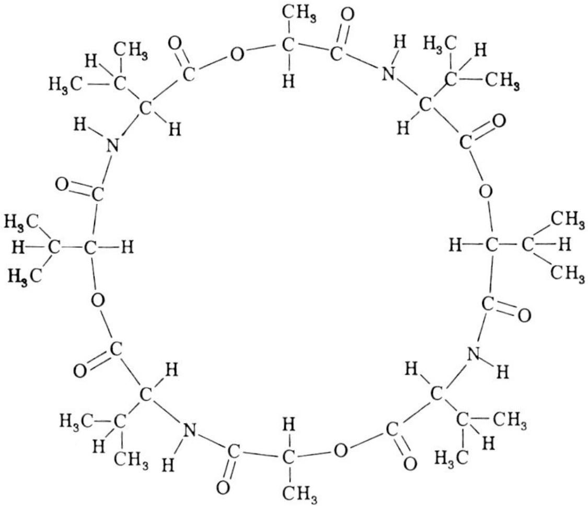

    Of the four valine residues in this structure, two are of the L-configuration and two of the D-configuration.

POSTSCRIPT

Since this book reached the page proof stage, the point at which no further major changes in the text can be made, some very important developments have taken place. The only possible solution to the dilemma of what to do about these developments was to write a postscript in which these developments could be summarized in sufficient detail so that the reader could rewrite the relevant sections of the book on his own.

The study of the energy cycle in mitochondria *in situ* by R. A. Harris *et al.* has established that there are only four configurational states—two of which are nonenergized and two of which are energized. The two nonenergized configurations are the orthodox and the aggregated; the two energized configurations are the energized and energized-twisted. There are two possible configurational cycles: orthodox → energized → energized-twisted → orthodox; or aggregated → energized → energized-twisted → aggregated. Mitochondria *in situ* usually undergo the first cycle whereas isolated mitochondria in 0.25 *M* sucrose usually undergo the second cycle. But by manipulation of conditions, mitochondria *in situ* can undergo the second cycle and isolated mitochondria can undergo the first cycle.

In the orthodox configuration, the cristae are maximally contracted, i.e., the intracristal space is minimal and the matrix space is maximal. In the aggregated configuration, the cristae are maximally expanded, i.e., the intracristal space is maximal and the matrix space is minimal.

The generation of the aggregated conformation sets the stage for engagement via the energizing processes since the distended cristae in this configuration are now apposed against their equally distended neighbors. The fact that there are only four possible configurational states in all mitochondria whether isolated or *in situ* leads to a great simplification of the configurational picture.

E. F. Korman has established the basic principle that the topology of the mitochondrion is invariant throughout the entire configurational cycle. None of the rearrangements in membrane form affect this topology, and this includes the phenomenon of tubularization which leads to the formation of tubules in the energized-twisted configuration. Tubes are formed by invaginations from the cristael membrane in the energized configuration, not by comminution as we originally postulated. It is this constancy of topology which explains why mitochondria will swell to the point where the cristae disappear, and then shrink back to the original size with all the cristae neatly back in place, why the mitochondrion in the energized-twisted configuration can easily revert back to the aggregated or orthodox configuration when the cristael membrane is discharged.

The transition of the cristael membrane from the orthodox to the aggregated configuration is a reversible one and is an important cellular control point. Allmann *et al.* have shown that Ca^{2+} induces the orthodox while Mg^{2+} or Mn^{2+} induces the aggregated configuration. High osmotic pressure (e.g., a medium 0.25 M or higher in sucrose) will induce the aggregated configuration whereas a lower osmotic pressure will induce the orthodox configuration. By manipulation of the osmotic pressure and the level of Ca^{2+} and Mg^{2+}, it is possible at will to determine which configurational state will obtain under a given set of conditions. When the level of Ca^{2+} bound by the mitochondrion exceeds a critical amount, the mitochondrion can be frozen in the orthodox configuration in the sense that the energizing process will not compel the formation of the next configurational stage, namely the energized configuration. Such configurationally frozen mitochondria are completely uncoupled. This points up dramatically that engagement of cristae is a *sine qua non* for the consummation of coupling.

The transition of cristae in the orthodox to cristae in the aggregated configuration can be controlled not only by the $Ca^{2+}:Mg^{2+}$ ratio and by the osmotic pressure but also by appropriate drugs, anesthetics, endotoxins, and possibly hormones. The freezing of the mitochondrion in the orthodox configuration is probably of great physiological significance in hibernating animals and in tissues in which heat production is a normal process.

The study of uncouplers by M. Lee and R. A. Harris has revealed that there is a wide variety of mechanisms by which uncoupling in its widest sense can be achieved. The prime uncouplers such as 2,4-dinitrophenol and *m*-chlorocarbonylcyanide phenylhydrazone suppress the nonenergized to energized conformational transition; the other uncouplers intervene in one or other of the two subsequent conformational transitions—energized to energized-twisted and energized-twisted to nonenergized. Fluorescein mercuric acetate, for example, suppresses the energized to energized-twisted transition by preventing the introduction of inorganic phosphate. Similarly, a medium of high osmotic pressure can suppress the energized to energized-twisted transition but not the transitions before and after. The uncouplers appear to be largely reagents which interact with and modify sulfhydryl groups—a token that sulfhydryl groups are playing a key role in conformational transitions.

The question of the localization of the citric cycle enzymes may at long last be resolved by the discovery by E. F. Korman and T. Wakabayashi that in the space between the boundary membranes and in the continuation of that space within the cristae (the intracristael space) there is an organized, paracrystalline structure which has a periodicity determined by the basepieces to which this structure is hinged presumably by electrostatic attraction. The tendency of the solubilized citric cycle enzymes to form such paracrystalline structures when exposed to polyanions such as phosphotungstate has been documented by J. Smoly and T. Wakabayashi. The outer and inner boundary membranes can be fully divested of any trace of citric cycle enzymes although in crude form both these membranes are intimately linked electrostatically to citric cycle enzymes. The question of the localization of the atractyloside barrier now has to be reopened. Is it in the paracrystalline structure with which the citric cycle enzymes are associated or in the outer boundary membrane?

The localization of ordered structures within the space between the two boundary membranes and within the lumen of the cristae throws new light on the structural organization of the mitochondrion, and particularly on the problem of mitochondrial biogenesis.

The essential features of the various transducing systems discussed in Chapter VIII have been presented in the form of naive models. However, the inexorable tide of progress in biochemistry has been such that since the preparation of our original manuscript some of our models, even considered as extreme simplifications, have come to require modification.

A good example of the development of new insights has been in the field of muscle biochemistry. It is now apparent from the elegant studies of H. E. Huxley that the distance between actin and myosin filaments does not necessarily remain constant during contraction. This is not consistent with the model which we have presented (cf. pages 127–131). It now seems that H-meromyosin is composed of a globular region of about 50×200 Å (probably composed of 2 subunits) with a rigid helical "tail" about 400 Å long. This rodlike tail is attached to the main shaft of the myosin fiber by a flexible "hinge"; a similar hinge attaches the globular region to the other end of the rod. The angle between the rodlike tail of the H-meromyosin and the shaft of the filament is an acute one, in the direction of the polarization of the fiber referred to previously.

This system permits a constant orientation of the globular region of H-meromyosin with respect to the actin fiber over a wide range of interfiber distance. As the distance between the actin and the myosin fibers increases, the angle between the rod and the myosin increases, and this is compensated for by a corresponding change at the hinge between the rod and the globular region.

An intriguing additional feature of this system is that once attachment is made between the globular region and a G-actin subunit, movement of the fibers relative to one another can be achieved by a tilt of the globular unit, i.e., by an energized change in the angle of the hinge at the globular end of the rod. Such a tilt could be achieved, for example, by a movement of the two subunits of the globular region with respect to one another.

How the globular units in fact attach to the actin fiber, how any tilt is coupled to the hydrolysis of ATP, and how such movement is controlled by Ca^{2+} through the mediation of additional protein factors now become open questions. What is certain is that the model which we have presented, involving helix-coil transitions of single polypeptide chains, is now obsolete. The basic principle remains of ATP-energized conformational rearrangement as the basis of muscular contraction.

REFERENCES

Harris, R. A., Williams, C. H., Caldwell, M., Green, D. E., and Valdivia, E., *Science* **165**, 700 (1969): on the energized configurations of heart mitochondria *in situ*.

Korman, E. F., unpublished studies: on the topology of the mitochondrion.

Allmann, D. W., Munroe, J., Wakabayashi, T., Harris, R. A., and Green, D. E., *Arch. Biochem. Biophys.* in press (1970): on the determinants of the orthodox to aggregated transition in adrenal cortex mitochondria.

Lee, M., Harris, R. A., and Green, D. E., *Biochem. Biophys. Res. Commun.* in press (1969): on the uncoupling action of fluorescein mercuric acetate.

Harris, R. A., Jolly, W., Williams, C., and Green, D. E., *Arch. Biochem. Biophys.* in press (1970): on the suppression of the second configurational transition by media of high osmotic pressure.

Korman, E. F., and Wakabayashi, T., unpublished studies: on the presence of ordered paracrystalline structures in the intracristal space and in the space between the boundary membranes of the mitochondrion.

Smoly, J., Addink, A., and Wakabayashi, T., unpublished studies: on the formation of paracrystalline states by the complex of soluble mitochondrial enzymes.

Huxley, H. E., *Science* **164**, 1356 (1969): on the mechanism of muscular contraction.

SUBJECT INDEX

Italic page numbers refer to formulas.

A

Actin, *see* Actomyosin
Actomyosin
  atypicality in respect to membrane modality, 124
  mechanism of contraction, 129, 198
  relative movements in, 125, 127, 128
Antimycin, 80, 91, 103, *174*
ATP(ADP), *174*
  energy transformations and, 2, 21, 81, 85, 87, 95, 102, 103, 106
  synthesis
    in chloroplast, 124, 139
    in mitochondria, 14, 19, 20
    in particles, 110
    at substrate level, 12, 115
  transfer of phosphoryl residue, 70, 110
ATPase complex, 51, 107
Atractyloside, 62, 102, *175*
Azide, 103, *175*

B

Basepiece, *see* Repeating units
Boundary membranes of mitochondria, 46, 48, 96
  enzyme activities, 52
  vectorial character, 54

C

Calcium ions in actomyosin contraction, 129, *see also* Translocation, Transport
Carnitine, 55, 62, *175*
Carotenoid, 135
Cell division, 149, 150
Chemiosmotic hypothesis, 84
Chlorophylls, 122, 135, *176*
Chloroplast
  appearance, 132, 134
  configurational changes, 140
  membranes, 133
  photosynthetic apparatus, 121, 122, 132, 133–140

201